Condensation Particle Counting Technology and Its Applications

Condensation Particle Counting Technology and Its Applications introduces the principles, key components, calibration methods, and applications of condensation particle counting systems.

This book delves into the adverse effects of fine particles on human health, along with the existing detection technologies for these particles. It discusses the theories and methods of optical measurement for fine particles and elaborates on the applicable conditions for each light scattering theory, providing a theoretical foundation for detecting particle-scattered light. This book concludes with an overview of the challenges posed by ongoing climate change and future research prospects in condensation particle counting (CPC).

This book is intended for industry professionals and environmental researchers specializing in particle and aerosol measurement, detection methods, and technology.

Condensation Particle Counting Technology and Its Applications

Edited by
Longfei Chen, Xiaoyan Ma, Guangze Li,
and Liuyong Chang

CRC Press is an imprint of the
Taylor & Francis Group, an **informa** business

Designed cover image: Longfei Chen

First edition published 2025
by CRC Press
2385 NW Executive Center Drive, Suite 320, Boca Raton FL 33431

and by CRC Press
4 Park Square, Milton Park, Abingdon, Oxon, OX14 4RN

CRC Press is an imprint of Taylor & Francis Group, LLC

Library of Congress Cataloging-in-Publication Data
Names: Chen, Longfei (Engineering professor), editor.
Title: Condensation particle counting technology and its applications /
edited by Longfei Chen, Xiaoyan Ma, Guangze Li, and Liuyong Chang.
Description: First edition. | Boca Raton, FL : CRC Press, 2025. |
Includes bibliographical references and index.
Identifiers: LCCN 2024009522 (print) | LCCN 2024009523 (ebook) |
ISBN 9781032729503 (hbk) | ISBN 9781032729510 (pbk) | ISBN 9781003423195 (ebk)
Subjects: LCSH: Air pollution–Measurement. | Particles–Measurement. |
Atmosphere–Measurement. | Dust–Environmental aspects. | Nanoparticles–Health aspects.
Classification: LCC TD884.5 .C657 2025 (print) | LCC TD884.5 (ebook) |
DDC 628.5/30287–dc23/eng/20240326
LC record available at https://lccn.loc.gov/2024009522
LC ebook record available at https://lccn.loc.gov/2024009523

ISBN: 978-1-032-72950-3 (hbk)
ISBN: 978-1-032-72951-0 (pbk)
ISBN: 978-1-003-42319-5 (ebk)

DOI: 10.1201/9781003423195

Typeset in Times
by codeMantra

Contents

Editors

Longfei Chen is a full professor in the School of Energy and Power Engineering at Beihang University, China. He obtained his bachelor's and master's degrees in Automotive Engineering from Tsinghua University and completed his Ph.D. in Engineering Science at the University of Oxford in 2010. His research focuses on particle emissions, spray and combustion, and ice nucleation in the atmosphere. Specifically, his work includes (1) developing measurement systems for sub-micron particle emissions, (2) studying heat and mass transfer in multiphase flows, and (3) investigating condensation and ice nucleation of atmospheric particles. He serves as the Secretary General of the Aviation Internal Combustion Engine Branch of the Chinese Society for Internal Combustion Engines and as a member of the SAE E31 Working Group of the ICAO Emission Standards Committee. His accolades include the First Prize in Technological Innovation awarded by the Chinese Society for IC Engines (2023) and the Chinese Society for Particuology (2023), the National Excellent Young Scholar Award (2019), the China Internal Combustion Engine Society Outstanding Researcher Award (2019), the 4th China-France Team Cooperation Innovation Award (R&D Award, 2017), and the Beijing Science and Technology Nova Star Award (2017).

Dr. Xiaoyan Ma is an associate professor at the Hangzhou International Innovation Institute of Beihang University. She obtained an interdisciplinary doctoral degree in Thermodynamics and Civil Engineering from the University of Paris-Saclay in 2020 and worked at Centre National de la Recherche Scientifique (CNRS) in France. Her research interests cover heat and mass transfer in porous media, multiphase flow, energy storage, and morphology and mechanisms of particulate matters. She has participated in national projects of France and international projects of the European Union. She has published more than 20 SCI-index journal papers and one book as a chapter contributor. She is leading a recent research project on combustion detection and fine particle identification supported by the Beijing Natural Science Foundation.

Dr. Guangze Li is an associate professor at the Hangzhou International Innovation Institute, Beihang University, China. He received his Ph.D. in Thermal Engineering from Beihang University in 2022 and has participated in a joint Ph.D. program at the National University of Singapore. His research focuses on the heat and mass transfer of multiphase flow, combustion, and emissions. He has published more than 20 papers in the top SCI-indexed journals, including *Energy, Fuel, Environmental Science & Technology*, and obtained financial sponsorship from the National Natural Science Foundation of China.

Dr. Liuyong Chang is an associate professor at the Hangzhou International Innovation Institute of Beihang University, China. He received his Ph.D. in Measurement Technology and Instrument from Beihang University in 2021. His research interests include particle emissions, nanoparticle detection, heterogeneous condensation, combustion monitoring, and combustion instability. In particular, he developed a measurement system for nanoparticles emitted from motor vehicles that meet China National VI emission standards. He has published nine refereed journal articles as the first or corresponding author, and obtained financial sponsorship from the National Natural Science Foundation of China.

Contributors

Liuyong Chang
Hangzhou International Innovation Institute
Beihang University
Hangzhou, China

Longfei Chen
Hangzhou International Innovation Institute
Beihang University
Hangzhou, China
and
School of Energy and Power Engineering
Beihang University
Beijing, China

Boxuan Cui
School of Energy and Power Engineering
Beihang University
Beijing, China

Xuehuan Hu
School of Energy and Power Engineering
Beihang University
Beijing, China

Guangze Li
Hangzhou International Innovation Institute
Beihang University
Hangzhou, China

Zhirong Liang
Hangzhou International Innovation Institute
Beihang University
Hangzhou, China

Lei Liu
Hangzhou International Innovation Institute
Beihang University
Hangzhou, China

Xiaoyan Ma
Hangzhou International Innovation Institute
Beihang University
Hangzhou, China

Kang Pan
Hangzhou International Innovation Institute
Beihang University
Hangzhou, China

Shanshan Tang
Hangzhou International Innovation Institute
Beihang University
Hangzhou, China

Jingsha Xu
Hangzhou International Innovation Institute
Beihang University
Hangzhou, China

Zheng Xu
Hangzhou International Innovation Institute
Beihang University
Hangzhou, China

Bin Zhang
Hangzhou International Innovation Institute
Beihang University
Hangzhou, China

Zichen Zhang
Hangzhou International Innovation Institute
Beihang University
Hangzhou, China

Shenghui Zhong
Hangzhou International Innovation Institute
Beihang University
Hangzhou, China

Meiyin Zhu
Hangzhou International Innovation Institute
Beihang University
Hangzhou, China

1 Overview of Fine Particle Number Concentration Measurement Theory and Technology

Meiyin Zhu, Bin Zhang, and Longfei Chen

1.1 INTRODUCTION

Rapid economic and societal growth coupled with advancing industrialization has led to a significant and consequential environmental pollution issue [1,2]. Among various environmental pollutions, air pollution has the broadest impact and poses a wide range of harms that directly affect human life [3,4]. Fine particulate matter ($PM_{2.5}$), as one of the major atmospheric pollutants, refers to airborne particles with an aerodynamic diameter equal to or smaller than 2.5 μm. Its sources are multifaceted, encompassing anthropogenic and natural origins [1,5]. Anthropogenic sources stem from human activities, including emissions from industrial processes like coal-fired power generation and steel smelting, as well as domestic sources like kitchen fumes and coal-fired heating, in addition to exhaust emissions from transportation industries such as motor vehicles and aircraft engines [1]. Natural sources arise from natural phenomena like soil and rock weathering, forest fires, volcanic eruptions, plant pollen, and ocean spray [6]. Despite the relatively low concentration of fine particulate matter in the Earth's atmosphere, its widespread distribution and intricate physical and chemical transformations are noteworthy. Moreover, the elevation of $PM_{2.5}$ concentration contributes to the formation of haze, which detrimentally impacts air quality, human health, global climate, and other aspects [1].

Particle detection can be broadly classified into two categories: mass concentration detection and number concentration detection. Mass concentration detection methods are well established and include manual monitoring gravimetric method, automatic monitoring gravimetric method, and beta ray method. Several standards have been formulated for the detection of particle mass concentration, such as "Technical Requirements and Detection Methods for Ambient Air Particulate Matter (PM_{10} and $PM_{2.5}$) Samplers" and "Technical Requirements and Detection Methods for Orifice Flowmeters for Calibration of Total Suspended Particulate Matter Sampling". On the contrary, the detection standard for particle number concentration started relatively late, and it was only around 2010 that the importance of measuring particle number concentration was gradually recognized. The current trend is to shift from measuring particle mass concentration to accurately measuring

DOI: 10.1201/9781003423195-1

particle number concentration. This is because nanoscale fine particles pose significant health hazards, and measurement methods based on particle mass cannot accurately represent their hazards due to the extremely small mass of nanoscale particles. On the contrary, measurement methods based on particle numbers have high sensitivity and can accurately assess the hazards of nanoscale particles. In the automotive and aviation sectors, regulations and standards for measuring the number concentration of nanoparticles (from "μg/m^3" to "#/mL") were introduced earlier to effectively assess the level of nanoparticle hazards originating from high-temperature sources. Furthermore, to mitigate the impact of particulate matter on the atmospheric environment, countries worldwide have implemented a series of particulate matter detection standards to evaluate ambient air quality, such as PM_{10} and $PM_{2.5}$ standards.

In the 1960s and 1970s, developed European countries and the United States took the lead in investigating the chemical composition and pollution sources of fine particles. Subsequently, the research in the field of fine particle measurement was further expanded to the aspects of spatiotemporal distribution, emission inventories, particle size distributions, and so on. From the 1980s to the 1990s, researchers began to study the health and pathological effects of fine particles from an epidemiological perspective. In recent years, the issue of global climate change has become increasingly prominent, leading to a growing research focus on the impact of fine particles on atmospheric chemistry and the global radiation balance.

The existing measurement technologies of fine particles mainly include filter sampling method and online semi-continuous monitoring method:

1. Filter sampling method:
2. The filter sampling method refers to the method of collecting fine particles on filters and conducting measurements/analyses in the laboratory. Factors such as temperature, humidity, and the type of filters will affect the measurement accuracy in the sampling process. The filter sampling method can usually be divided into dissolving system and non-dissolving system. The sampling process of the non-dissolving system is relatively simple, and the cost is relatively low. However, there are more potential sampling errors (such as particle loss and sample pollution) in the sampling process of the non-dissolving system. Nevertheless, considering the simple operation and low cost, the non-dissolving system is still widely used, especially in developing countries.
3. Online semi-continuous monitoring method:
4. Although the filter sampling method may be associated with multiple errors, its related techniques are mature. In order to overcome the disadvantages of the filter sampling method, such as time-consuming and low resolution, a variety of online semi-continuous monitoring methods have emerged. The presence and development of online semi-continuous monitoring methods have greatly promoted the understanding of fine particles. Although the online semi-continuous monitoring method greatly reduces the measurement time of fine particles and improves the detection resolution, the measurement accuracy suffers from great uncertainty. Therefore,

many researchers verified the measurement accuracy of fine particles by comparing the measurement results of the online semi-continuous monitoring method and the filter sampling method.

The number and size of fine particles are two important physical parameters for describing their properties. At present, the measurement of particle size distribution and number concentration of fine particles mainly includes two steps: particle size screening and counting. The dominant methods of fine particle size screening include electric-field-based screening, inertial-force-based screening, and centrifugal-force-based screening. The main methods for counting fine particles include electrostatic counting technology and condensation particle counting (CPC) technology. CPC is the most widely used method for counting fine particles [7–11]. At present, the vast majority of measurement equipment for the size distribution and number concentration of fine particles cannot be used directly for high-temperature sources. For the measurement of fine particles from high-temperature sources, the traditional method usually includes cooling and diluting before measuring the particles. However, the coupling relationship between homogeneous nucleation and heterogeneous nucleation of particles and the particle loss mechanism are complex and changeable, which leads to low measurement accuracy of high-temperature measurement.

1.2 SOURCES AND IMPACTS OF FINE PARTICULATE MATTER

1.2.1 Sources of Fine Particulate Matter

Atmospheric particulate matter refers to the complex mixture of solid and liquid particles suspended in the air. It can be classified into coarse particulate matter (2.5–10 μm), fine particulate matter (≤2.5 μm), and ultrafine particulate matter (≤0.1 μm) based on their aerodynamic diameters. Fine particulate matter originates from diverse anthropogenic and natural sources, including industrial emissions, transportation exhaust, household emissions, soil and rock weathering, forest fires, volcanic eruptions, plant pollen, and sea spray.

1.2.1.1 Anthropogenic Sources of Fine Particulate Matter

The industrial production process contributes to the release of fine particulate matter into the atmosphere. Two types of emission modes can be identified: organized emissions and unorganized emissions. Organized emissions primarily involve the collection of coal-fired flue gas and industrial dust from coal-fired industrial kilns during production, which are then discharged through pipes. Industrial kilns differ from industrial boilers as they mix coal and raw materials for high-temperature calcination, resulting in the emission of particulate matter components not only from coal but also from raw materials. Unorganized emissions, on the other hand, pertain to the particulate matter generated during the transportation, crushing, and extrusion of raw materials or products, as well as the release of pollutants through solvent volatilization in industrial production processes. Additionally, gaseous pollutants such as sulfur dioxide, nitrogen oxides, and volatile organic compounds produced during industrial production can undergo secondary oxidation reactions in the atmosphere, leading to the formation of fine particulate matter. Hence, these gaseous pollutants serve as significant precursors for $PM_{2.5}$ formation.

Vehicular exhaust emissions are a significant contributor to atmospheric fine particulate matter. As the number of motor vehicles continues to increase, the exhaust emissions from vehicles have become a major source of particulate pollution, prompting governments and researchers worldwide to pay close attention to this issue [12]. Taking China as an example, according to the "China Mobile Source Environmental Management Annual Report" released by the Ministry of Environmental Protection in 2022, the number of motor vehicles in China reached 395 million in 2021, a 6.2% increase from 2020, resulting in a particulate matter emission of 1,557.7 tons [13]. Vehicular exhaust gases contain not only combustion particles but also various semi-volatile and volatile substances. These substances can undergo transformations from gaseous to liquid or solid particles under changing environmental conditions such as temperature and pressure. In the field of automotive engineering, vehicular exhaust emissions typically consist of three types of particles: nuclei, accumulation, and coarse mode particles. Nuclei mode particles, defined as particles with a size less than 50 nm ($dp < 50$ nm) [14], make up the majority, accounting for over 90% in number but approximately 1.2% in mass [15]. Accumulation mode particles, with sizes ranging from 50 to 500 nm, represent a smaller proportion in number but a larger proportion in mass. These particles are relatively stable and less affected by the ambient environment, although a small portion may re-volatilize during sampling and dilution. Effective reduction of accumulation mode particles can be achieved through particle filters [12]. Coarse mode particles, with diameters larger than 500 nm, constitute 5% to 20% of the total mass concentration [15]. They are not directly generated in the engine combustion chamber but are formed through the re-entrainment of particles deposited in the engine cylinder, exhaust system, and exhaust sampling system. These particles can easily detach and re-enter the exhaust gas due to the unstable nature of the automobile exhaust system caused by air flow impact. Although the emission of coarse mode particles is sporadic and depends on previously deposited particles, their impact on the number and volume concentration of particles emitted by the engine is minimal due to their low number concentration [12]. The particle size distribution characteristics of particulate matter emitted by typical motor vehicles are shown in Figure 1.1 [12].

Unlike traffic emissions, aviation emissions typically consist of smaller particle sizes and primarily occur at high altitudes. The exhaust gas emitted by aircraft engines contains a significant amount of fine particulate matter with aerodynamic diameters less than 100 nm. As the sole anthropogenic source of particle emissions in the atmosphere at cruising altitude, aviation emissions have a crucial impact on the atmospheric environment in airport areas and global climate change, attracting substantial attention from the international community [16,17]. While advancements in aero-engine technology have reduced emissions from individual engines, the overall global air pollutant emissions continue to increase due to the rapid growth in global air travel demand, expansion of airport capacity, and the substantial rise in air transport volume [18]. The pollutant emissions from aero-engines are influenced by the air-to-fuel ratio (AFR). When the AFR is low, the engine's combustion chamber

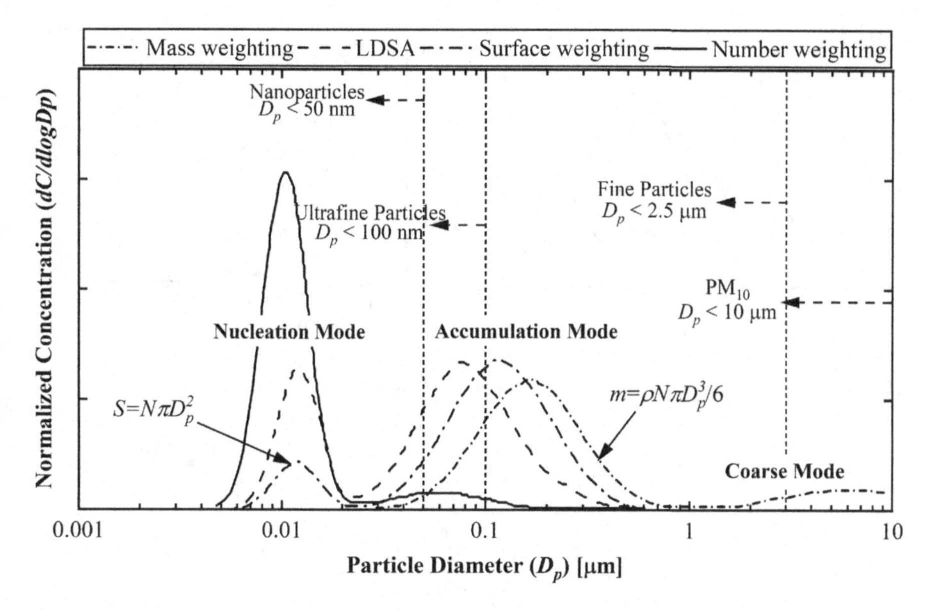

FIGURE 1.1 The particle size distribution characteristics of particulate matter.

operates in an oil-rich combustion state, leading to incomplete fuel combustion and the release of solid soot particles, unburned hydrocarbons (UHC), CO, and other pollutants. In a stoichiometric combustion state with sufficient air, the engine produces a significant amount of nitrogen oxides (NO_x) [19]. When the air volume is high, the engine operates in an oil-lean combustion state, resulting in reduced overall pollutant emissions. Ideally, exhaust gas should only contain CO_2 and SO_2 under ideal combustion conditions. However, in practice, due to incomplete fuel combustion and the use of lubricating oil, exhaust gas from modern civil aviation engines also contains gas components such as CO and NOx, as well as volatile organic compounds (VOCs) [20,21], including polycyclic aromatic hydrocarbons, and various particulate matter (PM). Aviation-emitted PM is a polydisperse mixture that can be categorized as volatile and non-volatile PM (nvPM) based on its volatility [22,23]. The Alternative Aviation Fuel Experiment (AAFEX) measurement experimental system conducted by NASA analyzed gas and particulate emissions from a CFM56-2C1 aero-engine and demonstrated the potential impact of using alternative aviation fuels on the atmospheric environment and climate change. The Scanning Mobility Particle Sizer (SMPS) results, shown in the right figure of Figure 1.2, revealed that the particulate emissions from the aero-engine exhibited two particle modes that varied with thrust conditions. The micro/nano nucleation mode particles ranged from 10 to 30 nm and consisted entirely of volatile substances, while the soot mode particles ranged from 30 to 200 nm and formed through the adsorption of volatile substances onto non-volatile soot components. Real-time aerosol mass spectrometry results, shown in the left figure of Figure 1.2, revealed that the volatile components in both particle modes comprised organic matter and sulfate, with minimal presence of nitrate components. These components influence particle moisture absorption, which is a crucial factor in

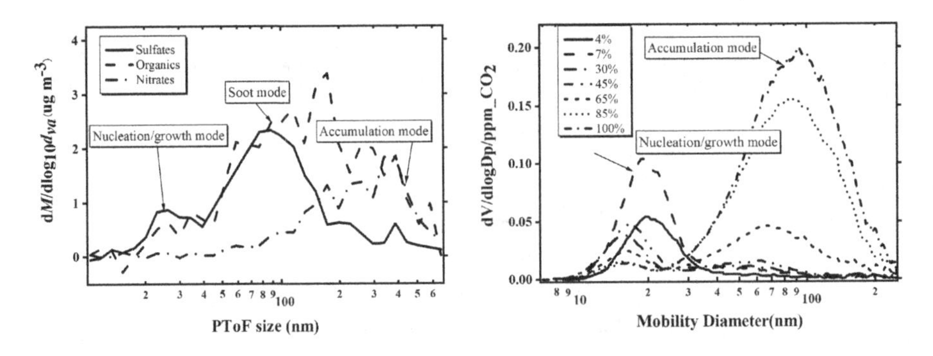

FIGURE 1.2 Chemical composition (left) and particle size distribution (right).

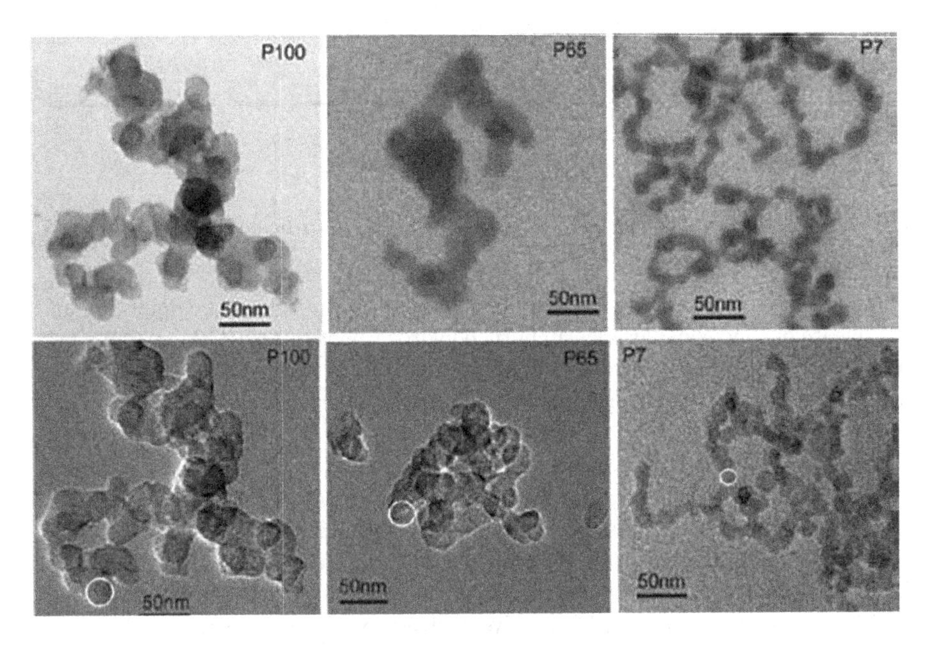

FIGURE 1.3 TEM observation diagrams of soot particles under 100%, 65%, and 7% power of an aero-engine [24].

determining the final particle size in the atmospheric environment. The nvPM emitted by aero-engines exhibits a typical fractal aggregate structure formed through the agglomeration of initial particles. The microstructure is highly complex, featuring branched, single-chain, spherical, or quasi-spherical forms [24]. Figure 1.3 displays a transmission electron microscope observation of nvPM with various morphologies collected at the nozzle of an aero-engine.

In addition to the aforementioned anthropogenic emissions, fine PM also originates from sources such as dust, the catering industry, fireworks and firecrackers, and agricultural straw combustion. These sources have been recognized as significant contributors to the presence of fine PM in the atmosphere [1,2]. Fugitive dust is a prominent open pollution source that occurs when dust on the ground is resuspended and enters the atmosphere. It constitutes a crucial portion of the total suspended particles in the ambient air [1]. The process of dust raising can be categorized into primary dust and secondary dust [5]. Primary dust arises from the air flow during the handling of bulk materials, causing dust to escape from material piles. Secondary dust, on the other hand, is generated by indoor air flow and ventilation, leading to the resuspension of settled dust on equipment, floors, and buildings [1,5]. Emissions from cooking activities in the catering industry contain a substantial amount of volatile organic compounds. These compounds can undergo further chemical reactions in the atmosphere, resulting in the formation of organic $PM_{2.5}$, also known as secondary organic aerosols [1,2,5]. Fireworks and firecrackers release gases such as CO_2, CO, SO_2, NO_x, and $PM_{2.5}$, among other pollutants, through a series of complex chemical reactions [5,25]. The combustion of crops, plants, and other biomass during agricultural production activities is another contributor to the formation of atmospheric fine particles. For instance, the burning of maize straw produces fine PM that is rich in elements like potassium, sulfur, and chlorine, as depicted in Figure 1.4 [26]. Biomass combustion generates a

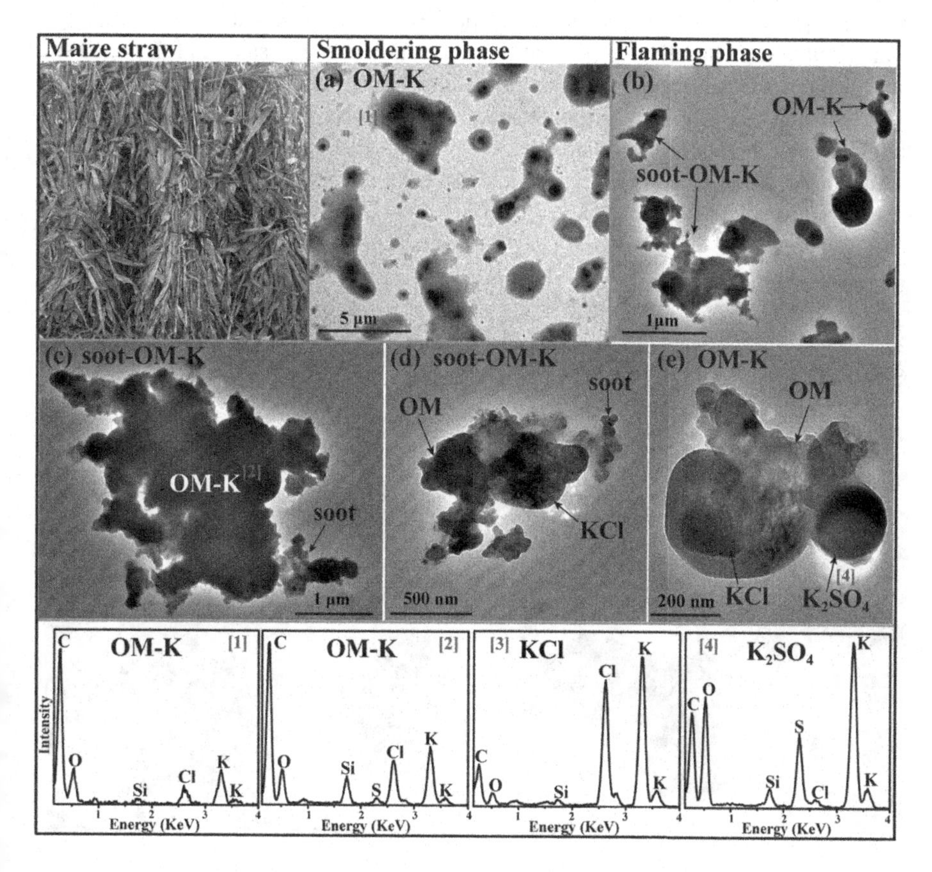

FIGURE 1.4 TEM image and EDS spectrum of single primary particle produced by corn straw combustion [26].

significant amount of smoke, which significantly reduces visibility, disrupts normal transportation, increases the risk of traffic accidents, and affects the smooth takeoff and landing of aircraft.

1.2.1.2 Natural Sources of Fine Particulate Matter

In addition to the fine PM generated by anthropogenic sources, a significant amount of fine PM is also produced by natural events and phenomena. Natural sources include natural soil fly ash, sea salt particles formed by splashing sea foam, plant pollen, spores, bacteria, and more, as shown in Figure 1.5 [27]. Natural events such as volcanic eruptions, forest fires, bare coal fires, and sandstorms can release substantial quantities of fine PM into the atmosphere. For instance, sea salt particles derived from the ocean contain elements like Na, Cl, and small amounts of Mg, Ca, K, and S. Scanning electron microscope (SEM) observations conducted by Laskin et al. demonstrated that sea salt particles generated by ocean waves in polluted coastal air ranged in size from 0.15 to 2 μm, with a median diameter of 1 μm. When sea salt from the ocean is transported to continental polluted air, it becomes an important reactant as it absorbs acidic gaseous substances such as SO_2, NO_x, and organic acids from the moist air [28–30]. Throughout the aging process, square sea salt particles undergo transformations, leading to the formation of partially aged sea salt and fully aged sea salt particles through heterogeneous reactions. These particles contain compounds such as $NaNO_3$, Na_2SO_4, or organic salts containing Na [31–33]. It is worth noting that the reaction between polluted air and sea salt aerosols from the ocean can result in increased surface ozone levels, which can have adverse effects on air quality in heavily polluted coastal areas [31].

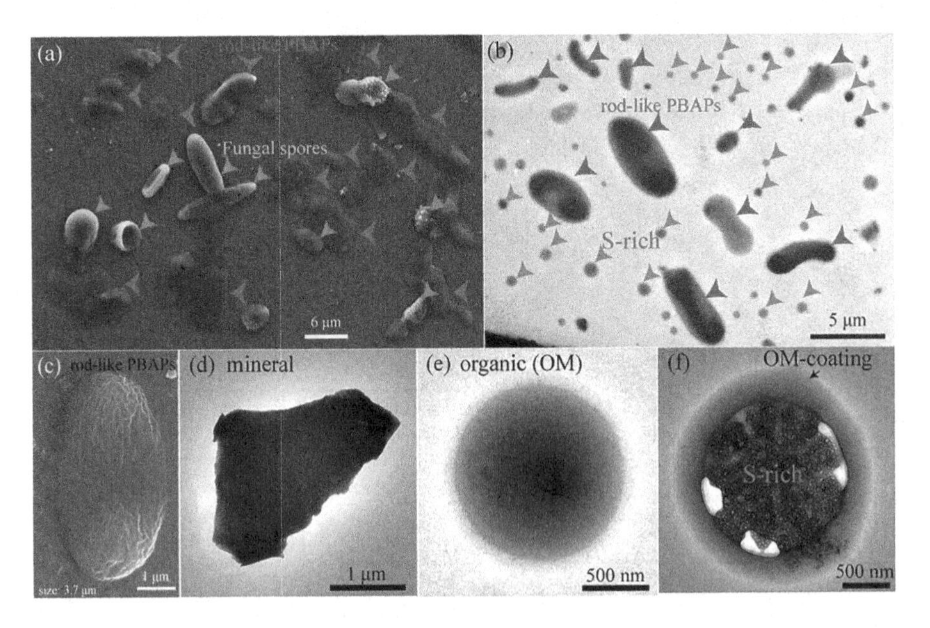

FIGURE 1.5 Low magnification SEM and TEM images of individual particles collected from the forest air [27].

1.2.2 The Impacts of Fine Particulate Matter on the Environment and Climate

The impact of atmospheric fine PM on the environment and climate is primarily manifested in several aspects, including ambient air quality, visibility, the water cycle, and climate change.

When atmospheric fine particles, which are rich in toxic substances and microorganisms, accumulate to a certain level, they can have a severe impact on local and even regional air quality, leading to air pollution issues and posing risks to human health. Air pollution is a significant global environmental problem that is closely tied to human activities. Although atmospheric fine PM is present in trace amounts in the atmosphere, it plays a crucial role in important environmental issues such as local and regional air pollution and global warming [1,2,5]. Fine particles in the atmosphere influence climate in two ways [34,35]. Firstly, they can directly absorb (e.g., soot) or scatter (e.g., sulfate) short-wave and long-wave solar radiation, affecting the radiative forcing of Earth's atmospheric system. Secondly, fine particles participate in cloud formation, evolution, and dissipation processes, altering cloud droplet size, microphysical structure, and optical properties, thereby impacting cloud lifetime and precipitation efficiency. This phenomenon is referred to as the indirect effect of PM. The indirect effects of fine particles can be further classified into two types: the cloud albedo effect and the cloud lifetime effect. The cloud albedo effect refers to the increase in cloud droplet concentration and decrease in cloud droplet radius caused by anthropogenic fine particles, resulting in changes in cloud radiation properties and an increase in cloud albedo. This effect is also known as the first indirect effect or the "Twomey effect". The cloud lifetime effect occurs when anthropogenic aerosols cause a reduction in cloud droplet size, adjusting liquid water content and cloud thickness, reducing precipitation efficiency, and prolonging cloud lifetime. This effect is also known as the second indirect effect or the "Albrecht effect". In addition to these indirect effects, fine particles also have a semi-direct effect on PM. This implies that fine particles with absorbing properties have the ability to absorb solar radiation, resulting in atmospheric heating. This can lead to increased stability in the lower atmosphere and even the potential evaporation of cloud droplets. According to the Seventh Assessment of the Fourth Assessment Report by the United Nations Intergovernmental Panel on Climate Change, the total direct radiative forcing of aerosols is $-(0.5 \pm 0.4)$ W/m^2.

1.2.3 Effects of Fine Particulate Matter on Human Health

In addition to its impact on the environment and climate, fine PM also significantly affects human health. $PM_{2.5}$, which represents a significant proportion of atmospheric particulate matter, poses a greater health risk due to its complex structure and larger surface area compared to larger particles. $PM_{2.5}$ has a higher toxicity and can readily adsorb harmful heavy metals and organic substances [36–40]. Numerous studies have demonstrated the substantial harm of fine PM to the respiratory system, nervous system, cardiovascular system, and reproductive system in humans [41–46]. According to estimates by the World Health Organization (WHO), 3% of cardiopulmonary diseases and 5% of lung cancers worldwide can be attributed to $PM_{2.5}$ exposure [47]. The International Agency for Research on Cancer has classified $PM_{2.5}$ as a Group 1 carcinogen. In 2015 alone, atmospheric fine PM pollution was responsible for 4.2 million deaths worldwide, accounting for 7.6% of total global deaths, with 59% of these deaths occurring in East and South Asia [40].

The harmful effects of atmospheric fine particles on human health primarily manifest in respiratory diseases. High concentrations of pollutants can lead to acute poisoning and even death [48]. Long-term exposure to low-concentration pollutants can result in bronchitis, bronchial asthma, emphysema, and lung cancer. The impact of PM on human health largely depends on the concentration of particles and the duration of exposure. Research data demonstrates a positive correlation between an increase in hospital visits due to upper respiratory tract infections, bronchitis, asthma, pneumonia, emphysema, and the concentration of particles in the atmosphere. The health risks associated with exposure to an environment where other pollutants and PM coexist are significantly more severe compared to exposure to a single pollutant [49].

The particle size of atmospheric PM is another significant factor that poses a threat to human health. Research by Kelly suggests that finer particles are less likely to deposit due to their small inertia and are more likely to be inhaled deeper into the human body [51]. Particles larger than 2.5 μm are typically blocked by the nasal cavity and remain in the upper respiratory tract, while particles ranging from 0.1 to 2.5 μm can enter the trachea and lower respiratory tract. Nanoparticles smaller than 0.1 μm can reach and deposit in the alveoli [52]. Figure 1.6 illustrates a schematic diagram of the effect of PM on the human respiratory system [53]. Smaller particle sizes correspond to larger specific surface areas and higher physical and chemical activity, exacerbating the occurrence and progression of physiological effects. Furthermore, the surfaces of fine particles can adsorb various harmful gases and pollutants in the air, including toxic metals, carcinogens, and pathogenic bacteria. Research results indicate that 60% to 90% of the harmful substances in PM exist in particles with an aerodynamic diameter of $\leq$10 μm [1]. Certain potentially toxic metal elements and organic compounds, such as lead, cadmium, nickel, manganese, and Polycyclic Aromatic Hydrocarbons (PAHs), are primarily attached to particles with a size of less than 2 μm [1], which can easily penetrate the deep parts of the human respiratory tract and be absorbed into the bloodstream through the capillary region of the alveoli, thereby causing harm to the human body.

1.3 PARTICLE MEASUREMENT STANDARD

1.3.1 Particle Mass Concentration Measurement Standard

For decades, the adverse health and climate effects of PM emitted from various combustion sources have been extensively studied [54–56]. In 2008, detailed scientific investigations were initiated in the United States and Europe to better understand and quantify the characteristics of particulate emissions [57–61]. Meanwhile, the first proposals to introduce International Civil Aviation Organization (ICAO) particulate

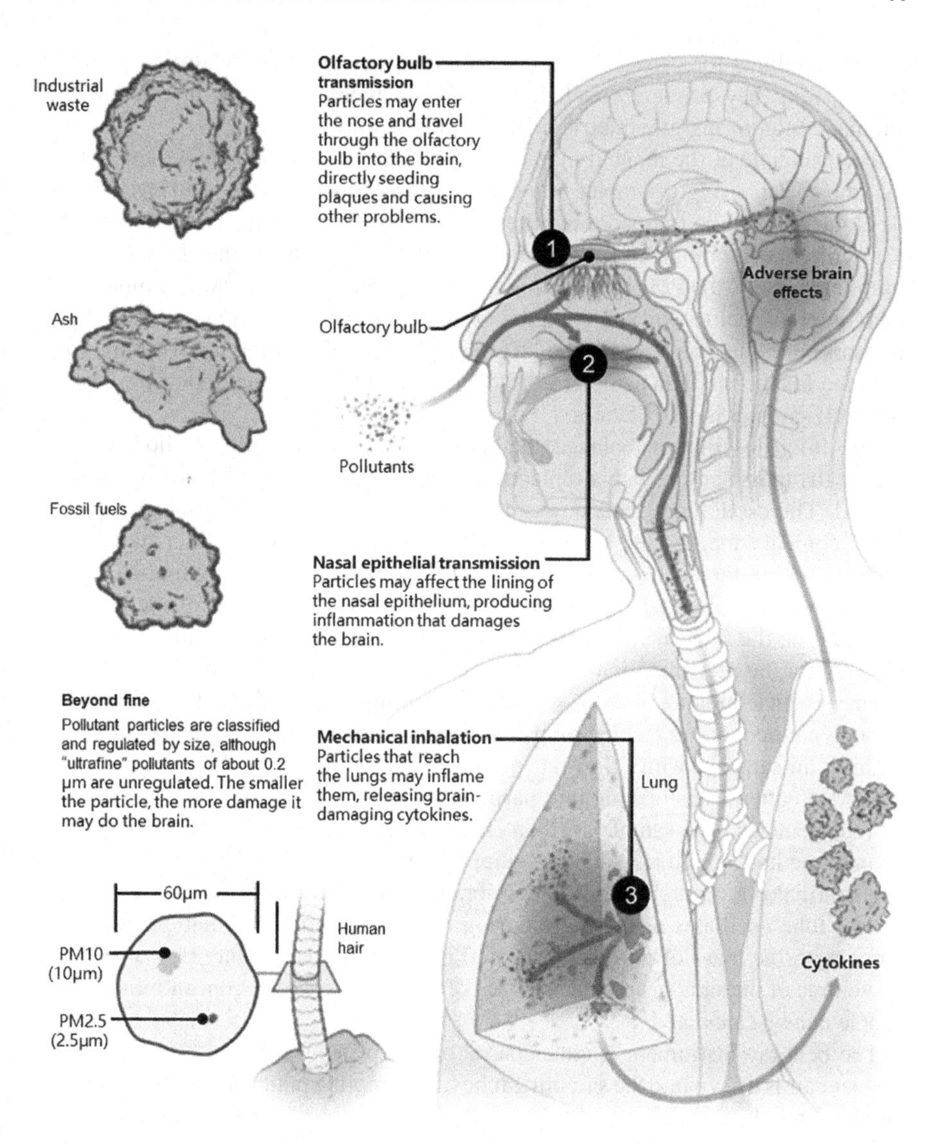

FIGURE 1.6 Schematic diagram of the effect of particulate matter on the human respiratory system [50].

standards for aircraft engines were introduced. Prior to 2010, the ICAO Committee on Aviation Environmental Protection (CAEP) primarily focused on formulating emission standards and recommended practices (SARPs) for PM based on smoke emissions [62]. The purpose of these SARPs was to mitigate the impact of aero-engine PM emissions on local air quality. In China, the "Ambient Air Quality Index Technical Regulations (Trial)" were implemented in 2016, providing a quantitative description of air quality conditions [63–66].

The standards for particle mass concentration detection are relatively abundant [67–70], and they primarily include the manual monitoring gravimetric method, automatic monitoring gravimetric method, and beta-ray method.

The principle of the manual monitoring gravimetric method is based on extracting ambient air using a sampler at a constant sampling flow rate, causing $PM_{2.5}$ in the ambient air to be trapped on a filter membrane. By measuring the mass change of the filter membrane before and after sampling and considering the cumulative sampling volume, the $PM_{2.5}$ concentration can be calculated. The $PM_{2.5}$ sampler does not require a specific working point flow rate. The typical conditions are as follows: the working point flow rate for the large sampler is $1.05\,m^3/min$, the working point flow rate for the medium sampler is 100 L/min, and the working point flow rate for the small sampler is 16.67 L/min.

The micro-oscillating balance method utilizes an oscillating hollow conical tube as the mass sensor, with a replaceable filter membrane installed at the oscillating end. The oscillation frequency is determined by the characteristics and quality of the conical tube. As the sampling airflow passes through the filter membrane, particles within the airflow are deposited on the membrane, leading to a change in the weight and, subsequently, the oscillation frequency. The mass of the particles deposited on the filter membrane is then calculated based on the frequency change, allowing for the determination of particle mass concentration. The micro-oscillating balance particle monitor consists of a PM_{10} sampling head, $PM_{2.5}$ cutter, filter membrane dynamic measurement system (FDMS), sampling pump, and instrument host. Ambient air samples, with a flow rate of $1\,m^3/h$, pass through the PM_{10} sampling head and $PM_{2.5}$ cutter, transforming into particulate matter sample air that meets technical requirements. The sample air then enters the main unit of the micro-oscillating balance method monitor, which is equipped with the FDMS. Within the unit, the sample air flows through the filter membrane, where PM is collected. The hollow conical tube oscillates reciprocally during operation, and its oscillation frequency varies with the mass of particles collected on the filter membrane. By considering the volume of the sample and the presence of these particles, the concentration of the sample can be calculated.

The beta-ray instrument operates based on the principle of beta-ray attenuation. Ambient air is drawn into the sampling tube by a sampling pump and discharged after passing through a filter membrane. PM is deposited on the filter membrane. When the beta ray passes through the filter membrane with deposited PM, the energy of the beta ray attenuates. By measuring this attenuation, the concentration of PM can be calculated. The beta-ray method particle monitor consists of a PM_{10} sampling head, $PM_{2.5}$ cutter, sample dynamic heating system, sampling pump, and instrument host. In the sample dynamic heating system, the relative humidity of the sample air is adjusted to below 35%. PM is collected on the filter membrane, which is automatically replaced after the sample enters the instrument's main unit. The instrument is equipped with a beta radiation source and detector, placed on opposite sides of the filter membrane. As the sample collection proceeds, more particles are collected on the filter membrane, increasing the mass of the particles. Consequently, the intensity of beta rays detected by the detector decreases. The detector's output signal directly reflects the mass change of the PM. The instrument analyzes the particle mass based on the detector data and combines it with the collected sample volume during the same period to calculate the particle concentration during the sampling period. Additionally, the instrument is equipped with a membrane dynamic measurement system, which allows for accurate measurement of PM volatilized during the process. This ensures that the final data can be effectively compensated and reasonably close to the true value.

1.3.2 Particle Number Concentration Measurement Standard

Currently, international standards for particle number concentration measurement primarily focus on vehicle emissions and aviation emissions [71–74]. Specifically, PM number-based metric standards related to motor vehicle emissions are predominantly utilized in China and the European Union. On the other hand, the measurement standards for the number concentration of PM associated with aviation emissions primarily rely on the relevant standards issued by the ICAO.

In 2020, China implemented the "Light Vehicle Pollutant Emission Limits and Measurement Methods (China Phase VI)", which provides clear guidelines for the detection of number and concentration in the field of motor vehicles. The following steps are required to be conducted before each test:

1. Activation of the particle-specific dilution system and measurement equipment, ensuring readiness of the sampling system.
2. Perform leak checks and zero-point checks:
3. Leak checking involves using a high-efficiency air filter at the inlet of the entire particle sampling system, with the measured concentration value by the particle number counter (PNC) being less than 0.5 particles/cm^3.
4. Zero-point check requires installing a high-efficiency air filter (HEPA) at the inlet of the particle sampling system. The measured value displayed by the PNC should be less than 0.2 particles/cm^3. After removing the HEPA filter, the measured value by the PNC should increase to at least 100 particles/cm^3. When the HEPA filter is reinstalled, the measured value by the PNC should be less than 0.2 particles/cm^3.
5. Confirm the correction function of the PNC and the volatile particle remover (VPR).

During the test, the following sampling steps should be followed:

1. Start the dilution system, sampling pump, and data acquisition system.
2. Initiate the PM and PN (particle number) sampling systems. The particle sampling system should continuously measure particle emissions, and the average concentration of particles is determined by integrating the measurement results throughout the test cycle. Sampling should begin before the engine is started or at the starting point, and it should be stopped at the end of the cycle.
3. Switch over the sampling.
4. Record the mileage of the dynamometer in each speed segment of the type I test cycle.

After the test, perform a gas analyzer check, bag analysis, and weighing of particle sampling filter paper.

Euro 7 is one of the European emission regulations that will be revised based on Euro 6 and is expected to be implemented in 2025 [75]. For M, N_1, and N_2 motor vehicles, both in the compression ignition and spark ignition stages, the maximum limit for particle mass concentration is 4.5 mg/km, and the PN concentration is 6.0×10^{11} #/km.

The ICAO, a specialized agency of the United Nations, is responsible for proposing new air transport policies and standardization innovations. Its subsidiary CAEP is responsible for formulating international standards and policies for aircraft engine emissions.

Before 2010, the CAEP primarily utilized smoke number (SN) as a measure to limit PM emissions in aircraft engines. The purpose of implementing the SN standard was to reduce the impact of PM emissions from aero-engines on local air quality. The SN measurement process involved measuring the reflectivity of a clean Whatman #4 filter membrane, passing a fixed mass of aero-engine exhaust gas (16.2 kg-gas/m^2) through the filter membrane, and then measuring the reflectivity of the filter membrane after the gas filtration. The change in reflectivity was used to calculate the SN value. However, as technology advanced, modern aero-engines achieved extremely low smoke emissions, often below the measurement error margin of $\pm$3SN. It is important to note that the SN method does not provide information on the number, composition, or particle size distribution of PM, making it inadequate for accurately assessing the environmental and health impacts of PM. Consequently, during the eighth CAEP meeting, it was acknowledged that in addition to regulating traditional pollutants, it was necessary to establish standards for limiting the quantity and quality of PM emissions.

According to the requirements of CAEP, the Society of Automotive Engineers (SAE) Committee on the Measurement of Air and Particulate Matter Emissions from Aircraft Engines (E31) took the lead in developing recommended sampling and measurement procedures for aero-engine nvPM (AIR 6241) and an aerospace recommended practice (ARP 6320). In 2016, at the tenth meeting of CAEP, the first regulatory standard for aero-engine nvPM emissions (CAEP/10) was adopted. This standard mandates that new aero-engines with a thrust greater than 26.7 kN, produced after January 1, 2020, must report nvPM emissions information. The maximum mass concentration limit in the CAEP/10 standard is determined based on the statistical relationship between nvPM mass concentration and SN. As a result, the restrictions on SN will be lifted in 2023.

The specific nvPM sampling measurement procedures are specified in Annex 16 (Environmental Protection) of the Convention on International Civil Aviation. The standard sampling system consists of three parts: the acquisition part, the transmission part, and the measurement part. The acquisition part includes a sampling probe and connecting pipelines. The sampling probe is positioned within a range of 0.5 times the diameter of the tail nozzle from the plane of the engine outlet. To ensure representative sampling, there should be at least 12 sampling points. The transmission part comprises a diluter, a cyclone separator, and connecting pipelines.

The temperature of the sampling gas at the diluter inlet should be maintained at 160°C ± 15°C. The dilution ratio of the diluter should range from 1:8 to 1:14, with pure nitrogen as the dilution gas. The temperature of the diluted gas should be kept at 60°C ± 15°C, and the flow rate should be maintained within the range of 25 ± 2 standard liters per minute (sLPM). The diluted gas then enters the measurement part through a constant temperature pipeline that is 24.5 ± 0.5 m long. To prevent instrument damage from clogging, a cyclone is installed to remove particles with aerodynamic diameters over 1 μm before the samples enter the measurement section. The measurement part primarily consists of two devices: an nvPM mass concentration detector and a number concentration detector. Before the sample gas enters the number concentration detector, it must pass through a VPR to eliminate interference from volatile compounds. The new nvPM standard will replace the SN standard and be officially implemented in 2023.

China's current implementation of China Civil Aviation Regulations Part 34—"Turbine Engine Aircraft Fuel Discharge and Exhaust Emission Regulations" still relies on the SN standard. However, it is anticipated that the ICAO benchmark for nvPM emissions will be introduced in the middle and late 2022, incorporating the latest standards and regulations.

1.4 CURRENT MEASUREMENT TECHNOLOGY OF FINE PARTICLES

1.4.1 Monitoring Network of Fine Particles

The United States, the European Union, and Canada have been monitoring fine particles in their respective networks for a considerable amount of time and have established long-term continuous monitoring capabilities.

1. The monitoring network for fine particles in the United States:
2. In the United States, there are two prominent monitoring networks that have the capability to monitor fine particles. These are IMPROVE (Interagency Monitoring of Protected Visual Environments) [76] and STN (Speciation Trends Network) [77]. In 1987, the National Park Service, the Fish and Wildlife Service, the Bureau of Land Management, the Forest Service, and the Environmental Protection Agency (EPA) proposed the project. The monitoring network in this project was originally established in Class I environmental areas designated by the US government. It primarily monitored aerosol composition and relevant visibility data. In 1999, IMPROVE was expanded to support the EPA's regional haze control efforts. Simultaneously, the EPA developed the STN to gain a better understanding of the composition of fine particles in the atmosphere and the disparities between urban and remote areas. The STN was developed as part of the Atmospheric Particulate Observatory Program in the United States. It was utilized to analyze the PM2.5 composition in urban areas and to identify regions that comply with newly established national environmental

air quality standards for particulate matter. The STN's PM2.5 measurement data can help identify PM sources and estimate their contribution to PM mass concentration. The Chemical Speciation Network was later established by STN and additional sites. The long-term operation of STN and IMPROVE has provided important data for studying the characteristics of fine PM in the US atmosphere.

3. The monitoring network of fine particles in the European Union:
4. The monitoring of fine particles in the European Union was initiated through the Cooperative Program for Monitoring and Evaluation of the Long-range Transmission of Air Pollutants in Europe (EMEP). The primary aim of the EMEP program is to provide governments and subsidiary bodies under the Convention on Long-Range Transboundary Air Pollution (LRTAP) Convention with reliable scientific information on a regular basis. This is to aid the development and review of international protocols on emission reductions, which are negotiated within the Convention [78]. The EMEP program has established a comprehensive monitoring network through various projects such as CREATE (Construction, use, and delivery of a European aerosol database) and GAW (Global Atmosphere Watch). Short-term monitoring programs like the ECRHS II (European Community Respiratory Health Survey Follow-up) have also been conducted by the European Union. The EMEP program primarily focuses on assessing acidification, eutrophication, and cross-border transport. However, it has also expanded to address other environmental concerns such as ground-level ozone formation, persistent organic pollutants, heavy metals, and fine particles.
5. The monitoring network of fine particles in Canada:
6. In Canada, the Canadian Air and Precipitation Monitoring Network (CAPMoN) has been utilized since 1983 to study regional patterns and trends of atmospheric pollutants, including acid rain, smog, PM, and mercury in both air and precipitation [79]. Operated by Environment and Climate Change Canada (ECCC), CAPMoN has replaced two older networks, the Canadian Precipitation Sampling Network (CANSAP) and the Air and Precipitation Network. Through CAPMoN, Canadians gain a better understanding of the sources and impacts of atmospheric pollutants. The valuable data provided by CAPMoN plays a significant role in monitoring environmental pollution in Canada.

The success of regional monitoring networks in the United States, European Union, and Canada has highlighted the importance of establishing similar facilities worldwide. The monitoring facilities for physical, chemical, and optical properties of aerosols are crucial in improving global climate change. However, China's fine particle pollution is more severe compared to the United States, European Union, and Canada, and the initiation of Chinese monitoring networks with fine particle monitoring capabilities was delayed. Since 2011, the Chinese Academy of Sciences has initiated the Campaign on Atmospheric Aerosol Research network of China (CARE-China), which has led to the establishment of the first comprehensive research platform for atmospheric aerosols [80]. This study focuses on the reduction of anthropogenic aerosol emissions in China, with the aim of combating climate change. The network, known as CARE-China, has been established based on the Chinese Ecosystem Research Network, with 36 sites equipped with standardized instruments, monitoring protocols, and data analysis specifications. Through the monitoring and research of atmospheric aerosols in China, CARE-China is able to provide continuous data on the mass concentration spectrum of PM2.5 in typical areas of the country. The data collected by CARE-China can be used to investigate the impacts of aerosols on climate and environmental changes and to develop localized climate models throughout China.

The majority of monitoring networks are limited to specific regions. However, there are two notable global monitoring networks: the GAW Programme and the Aerosol Robotic Network (AERONET) [75,81]. The GAW focuses on studying the changes and trends in atmospheric composition and associated physical parameters. One of the primary objectives of GAW is to facilitate assessments of the chemical composition of the atmosphere on a global level. By doing so, GAW provides accurate scientific data to policymakers at national and international levels. Additionally, GAW supports international conventions aimed at addressing the depletion of stratospheric ozone and monitors climate change and long-range transboundary air pollution. The GAW Programme offers information and services regarding atmospheric compositions to public sectors and decision-makers. This includes addressing the impact of steadily increasing amounts of greenhouse gases, particularly carbon dioxide, on the climate. Additionally, the program addresses the depletion of the protective stratospheric ozone layer, which has led to increased ultraviolet radiation and a higher incidence of skin cancer and other diseases. Lastly, the program also tackles the issue of urban air pollution, particularly fine particles, which has a negative impact on human health. The AERONET program, established by NASA and PHOTONS, is a federation of ground-based remote sensing aerosol networks that has been greatly expanded by networks such as RIMA, AeroSpan, AEROCAN, NEON, and CARSNET, as well as collaborators from national agencies, institutes, universities, individual scientists, and partners. For over 25 years, the project has provided a long-term, continuous, and publicly accessible database of aerosol optical, microphysical, and radiative properties, which has been instrumental in promoting aerosol research, validating satellite retrievals, and fostering synergies. The AERONET collaboration offers a wide range of observations including spectral aerosol optical depth, inversion products, and precipitable water across various aerosol regimes on a global scale.

1.4.2 Application of Advanced Detection Technology in Characterization of Fine Particles

During the 1960s and 1970s, developed countries took the lead in analyzing the chemical compositions and pollution sources of fine particles. This research expanded to include spatiotemporal distribution and physicochemical properties. Currently, fine particle characterization detection technology can be categorized as offline and online detection. Offline detection is known for its precision and sensitivity, but it is time-consuming. On the other hand, online detection is real-time and fast. While online detection cannot completely replace offline detection, it has brought a significant breakthrough to fine particle characterization.

Offline detection is commonly used in micromorphology observation, microstructure analysis, and chemical composition analysis.

a. In the field of micromorphology observation of fine particles, various instruments are utilized including SEM, helium ion microscope (HIM), atomic force microscope (AFM), scanning tunneling microscopy (STM), and others. SEM is capable of scanning fine particles by focusing on a high-energy electron beam, and the micromorphology can be acquired by collecting, amplifying, and processing the information that is carried by electrons (refer to Figure 1.7a). The HIM utilizes three different ion beams (gallium, neon, and helium) to achieve superior observation accuracy for fine particles compared to SEM (refer to Figure 1.7b). On the other hand, AFM can analyze the surface micromorphology of fine particles and provide three-dimensional characterization (refer to Figure 1.7c). In recent times, STM has found application in the characterization of micromorphology of fine particles. Its high-resolution

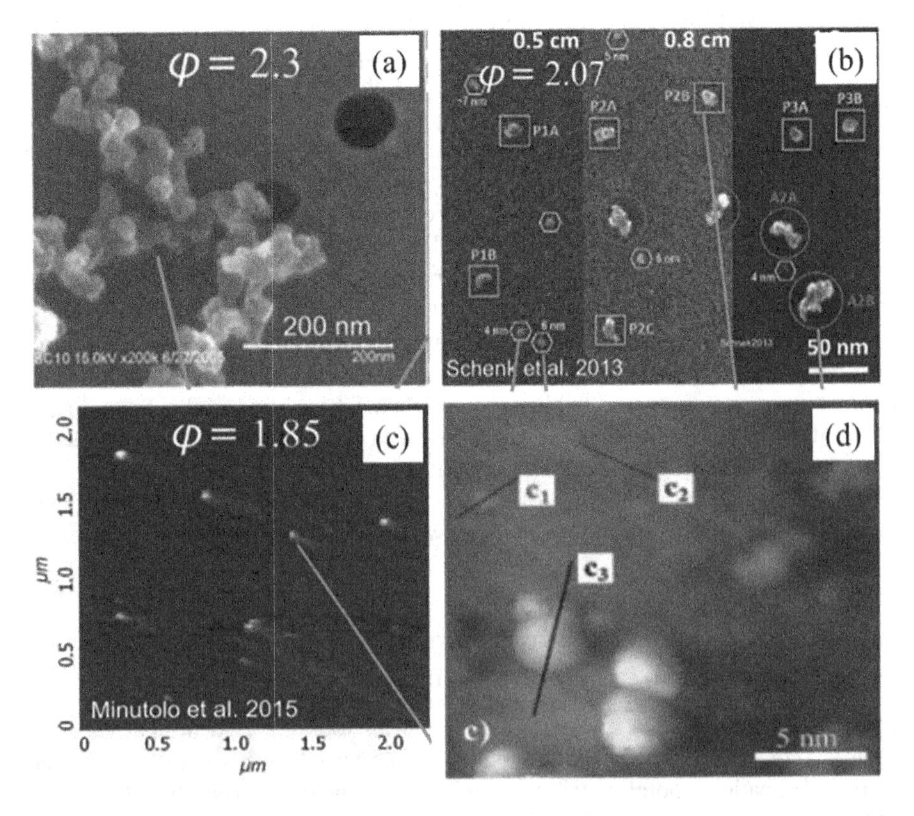

FIGURE 1.7 Micromorphology observation of fine particles: (a) SEM; (b) HIM; (c) AFM; (d) STM [82,83].

capabilities have the potential to provide more detailed information as illustrated in Figure 1.7d [82]. In addition to other methods, the use of an optical microscope is effective in observing the surface morphology and analyzing the particle size distribution of micron-sized fine particles.

b. In the field of microstructure analysis of fine particles, commonly used instruments include the transmission electron microscope (TEM) and AFM. Among them, TEM is one of the most effective instruments and is widely used. By using TEM, the internal microstructure of fine particles can be clearly observed, as shown in Figure 1.8a–c. On the other hand, high-resolution AFM is capable of scanning the microstructure of the fine particle surface, as shown in Figure 1.8d.
c. In the field of chemical composition analysis of fine particles, some of the major instruments used are Raman spectrometer, energy dispersive spectroscopy (EDS), X-ray photoelectron spectroscopy (XPS), and gas chromatography-mass spectrometer (GC-MS)[84,86–88]. Among these, the Raman spectrometer is particularly useful for obtaining information about molecular vibration and rotation by analyzing the scattering spectrum, making it a popular choice for determining material composition (Figure 1.9a). EDS is often used in conjunction with electron microscopy techniques like SEM and TEM, allowing for clear observation of

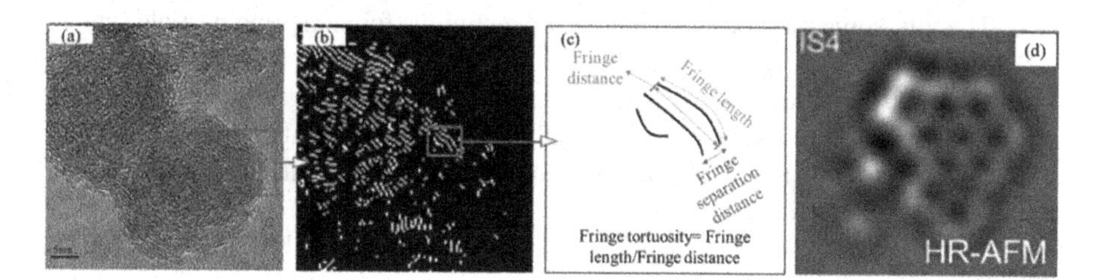

FIGURE 1.8 Microstructure analysis of fine particles: (a) the original TEM image; (b) internal microstructure; (c) microstructure feature extraction; (d) AFM image of surface microstructure [84,85].

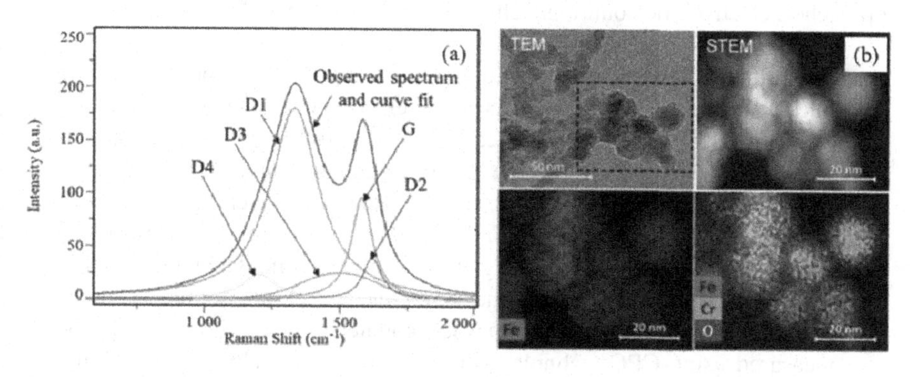

FIGURE 1.9 Chemical composition analysis of fine particles: (a) Raman spectrometer; (b) EDS [84,86].

fine particles and subsequent analysis of their chemical compositions (Figure 1.9b). XPS is commonly used for detecting the surface chemical compositions of small particles, offering a thorough chemical analysis with a detection depth of approximately 3–10 nm. On the other hand, GC-MS is a valuable tool that provides high selectivity, sensitivity, and resolution for identifying and quantifying complex compounds. It is particularly useful for determining trace fractions.

Online detection is a commonly used method to measure the particle size distribution and number concentration of fine particles.

Various techniques are employed for particle size distribution detection, including:

a. Laser scattering method is a suitable technique for measuring the particle size distribution of micron-sized fine particles in suspension, lotion, and powder samples. However, it may not be effective in measuring particles with a narrow particle size distribution due to low resolution.
b. The settling speed of particles in suspension can be used to measure particle size distribution. Using gravity or centrifugal force, larger particles settle faster than smaller ones, and this can be determined by applying Stokes' law, which states that the settling speed of spherical fine particles in water is proportional to the square of their diameter.
c. The ultrasonic method is a widely used technique for analyzing particle size distribution. As fine particles pass through the test area, sound waves are absorbed differently by particles of varying sizes, resulting in varying levels of attenuation at the receiving end. By analyzing the relationship between particle size and ultrasonic intensity attenuation, the size distribution of fine particles can be determined.
d. The identification of particle size can be accomplished by analyzing the negatively charged particles with a specific mass-to-charge ratio. The commonly used instrument for this purpose is the SMPS.

When it comes to detecting the concentration of fine particles, there are two primary approaches: electrostatic counting technology and CPC technology:

a. The principle of electrostatic counting technology is based on the fact that fine particles can generate an electric current after charging in the electrometer. This electric current value can be measured through the corresponding channel electrometer, and the PN concentration can be calculated based on the recorded values of electric current. However, the charged efficiency and charge of fine particles are influenced by their chemical compositions and physical morphology, making electrostatic counting technology highly dependent on the quality of the measured fine particles.
b. In addition to the use of optical particle counters, fine particles can also be measured using CPC technology. This method is widely recognized and specified by numerous emission regulations. Further details on this technology will be provided in Section 1.5.

1.5 CONDENSATION PARTICLE COUNTING TECHNOLOGY

CPC technology is commonly used to measure the quantity of fine particles. The primary tool for measuring the number concentration of fine particles is the condensation particle counter (CPC), which is developed based on this technology. Figure 1.10 depicts the typical structure of a CPC [89]. The technology behind CPC can be separated into two parts: hetero-coagulation technology which uses fine particles as condensation nuclei, and optical particle counting technology. The three main components of CPC are the saturator section, condenser section, and optics (optical particle counter (OPC)). The saturator section is responsible for providing the working fluid steam necessary for the hetero-coagulation process of fine particles. The condenser section has the ability to cause the working fluid steam to become supersaturated, leading to condensation on the surface of fine particles which promotes their growth. The OPC is capable of particle counting by detecting the Mie scattering pulse signal that is produced when particles pass through the light-sensitive area. The working process of CPC involves the flow of fine particles into the saturator section where they are exposed to the working fluid steam environment. The working fluid steam then reaches supersaturation in the condenser section of CPC and condenses on the particle surface. Fine particles are allowed to grow in size until they can be detected using optical methods. And the Mie scattered light pulse is then recorded as the particles flow through the OPC, allowing for the determination of the number concentration of particles.

CPCs, also known as condensation nucleus counters (CNCs), have a long history dating back to the late 19th century.

a. In early experiments, Coulier and Aitken used CPCs to study cloud condensation (1875, 1880) [90]. Aitken developed an expansion-type CNC in 1888, relying on adiabatic cooling to produce supersaturation. Wilson further advanced CNC technology in 1897 [91].

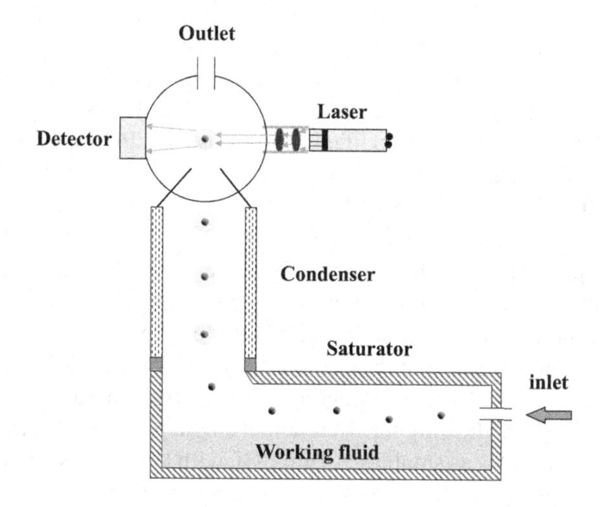

FIGURE 1.10 Typical CPC structure.

b. In the 1950s and 1960s, new CNC designs emerged for atmospheric research. Muhleisen and Holl introduced a steady-flow CNC in 1955 to study the relationship between atmospheric conditions and aerosol concentrations [92]. Burnashova and Kogan then demonstrated the first turbulent mixing-type CNC in 1960, which was further developed in later designs [91]. Up to 1975, CNCs relied on adiabatic expansion and valves to produce discontinuous flow condensation. The development of ultrafine condensation nucleus counters by Hoopes and Sinclair in 1975 and Bricard et al. in 1976 overcame this limitation [7].
c. In 1979, researchers in Scotland, Hogan, and Edinburgh delved into the origins of CNC. By 1980, Sem and Agarwal pioneered the TSI 3020 model, a continuous and single-particle counting laminar flow CNC [8]. A notable development occurred in 1982 when Kousaka introduced a CPC of the mixing type, inducing supersaturation through the swift blending of two flows—both saturated but at different temperatures. The year 1990 witnessed the unveiling of the ultrafine condensation particle counter (UCPC), a CPC variant tailored for detecting ultrafine particles [9]. Subsequent advancements in UCPC technology have empowered it to discern particles below the 2 nm threshold in contemporary times [10].
d. In 1991, Stolzenburg and McMurry pioneered a revolutionary UCPC using butanol as a base. This innovation allowed the measurement of particles down to a few nanometers, facilitated by a capillary sheath structure [11]. Before 1991, most conventional CPCs, particularly laminar flow CPCs, struggled to detect particles smaller than 4 nm. McMurry and Stolzenburg achieved a significant breakthrough with the invention of the first UCPC capable of detecting 3 nm particles [93]. By 1996, the TSI 3762, a commercially available CPC, entered the scene, distinguishing itself within its category with a notable high flow rate of 3 L/min [94].
e. In the year 2000, Dick et al. conducted a study on the Particle Height CPC (PH-CPC) TSI 3025, implementing modifications to the multichannel analyzer. The PH-CPC marked a significant milestone as the first instrument to undergo testing for its applicability. Subsequent developments in the field include the introduction of the water-based laminar flow CPC, TSI 3785, in 2003. In 2004, a series of modified butanol-based CPCs, namely TSI 3786 and TSI 3782, were introduced, complemented by models TSI 3775, TSI 3776, and TSI 3772 in 2005 [95].

Since Aitken's pioneering work in the late 19th century, CPCs have undergone significant development and commercialization [96]. Early CPC designs primarily relied on expansion cooling techniques and water as the working fluid. For many years, CPCs have been valued for their ability to detect particles below the size range accessible by optical methods. Commercial CPC instruments have been available since around 1970 [97], allowing the detection of single particles. Today, commercial CPCs can detect particles as small as 2.5 nm [98], with key factors determining their lower detection limit being the supersaturation ratio, working fluid, and nucleation temperature [99]. Modern CPCs typically utilize conductive cooling for particle detection due to its advantages.

Currently, there are several commercial CPCs available globally, including TSI in America, Grimm and Palas in Germany, which have achieved high levels of detection accuracy (with a lower limit of detection particle size of 3 nm). However, the development of commercial CPCs in China is still in its early stages. CPCs can be categorized based on the type of working fluid as either water-based or alcohol-based. Additionally, based on the mode of achieving supersaturation of the working fluid steam, CPCs can be classified into three categories: laminar flow CPC, adiabatic expansion CPC, and hybrid CPC. In the laminar flow CPC, the gas flow in the condenser section is laminar, leading to convection and diffusion heat transfer between the inner surface of the condenser and the working fluid steam. The different thermal and mass diffusion capacities of the working fluid cause the saturated steam to become oversaturated. Additionally, the laminar flow CPC can be categorized as either heating or cooling type based on the diffusion coefficients. In the adiabatic expansion CPC, gas with certain pressure and saturated working fluid particles are introduced into a closed chamber. The gas temperature is reduced through adiabatic expansion, causing the saturated steam to reach a supersaturated state. In the hybrid CPC, the supersaturated state of the working fluid steam is formed through turbulent mixing based on the temperature difference between the saturated working fluid steam and the gas to be measured. Current research on CPC focuses on the influence of particle properties on counting efficiency and the numerical model of heat and mass transfer in the condenser section. Wimmer et al. [100–105] tested the counting efficiency of various types of CPC under different parameters, while Giechaskiel et al. [106] investigated the influence of chemical components of particles on commercial butanol-based CPC and obtained the contact angle of different particles by comparing the counting efficiency predicted theoretically with experimental data. Additionally, a research team at Beihang University established a numerical model of CPC and revealed the influence of temperature changes in the saturator section and condenser section on the CPC counting efficiency through theoretical analysis and numerical simulation [107]. In a study conducted by the research team at Anhui Institute of Optics and Fine Mechanics, they utilized OPC to measure atmospheric aerosol particle count and compared it with particle quality monitoring data [108]. Hefei Institute of Material Science proposed a condensed-nucleus-based method to measure the number concentration of atmospheric fine particles [109]. Currently, one of the primary research focuses in the field of CPC is to investigate the minimum particle size that can be detected by the technology. Kangasluoma et al. [105] widened the temperature difference between the saturator and condenser sections to 36°C, and reduced the cut-off particle size from 10 to 3 nm. A research team at Tsinghua University has developed a fine particle condensation particle counter (UCPC) that uses diethylene glycol instead of the traditional working fluid, butanol. The UCPC was tested with NaCl particles ranging from 1 to 6 nm and was found to have a cut-off particle size of 1 nm [110,111].

Instruments developed using condensation nuclear particle counting technology are only capable of measuring the number concentration of fine particles at room temperatures. Direct counting of fine particles at high temperatures is not possible with this technology [112–115]. The conventional method for measuring fine particles from combustion sources involves cooling and diluting the sample before measuring the number concentration. However, this method has limitations due to the complex and variable coupling relationship between self-condensation and heterogeneous-condensation of particles caused by cooling, as well as particle loss mechanisms such as thermophoresis and diffusion electrophoresis. As a result, the accuracy of the number concentration measurement for fine particles using this method is low, with measurement errors for particles below 23 nm reaching up to 40%. For instance, this is evident in the case of measuring motor vehicle emissions [116,117].

In recent years, researchers have conducted numerous studies on direct counting technology for fine particles at high temperatures to improve the accuracy of measuring fine PN concentration:

1. Collings et al. from the University of Cambridge and Cambustion Ltd. Introduced the concept of a high-temperature condensation particle counter (HT-CPC, Figure 1.11) and conducted a series of experiments and simulation studies to evaluate its effectiveness [118]. The proposed HT-CPC by Collings et al. operates at high temperatures, causing the working fluid to evaporate into gas and limiting its ability to detect solid nanoparticles [118]. However, this system eliminates the need for complex preprocessing of gas samples and improves the accuracy of measuring fine PN concentration at high temperatures. Additionally, experimental verification of the nucleation properties of working fluids such as DEHS, flubiline, DC 704, Lesker 705, and Santovac 5 was conducted. The HT-CPC has a cutting size (d50) of less than 8 nm when using sodium chloride as the calibration particle, which refers to the particle size corresponding to the counting efficiency of 50%.

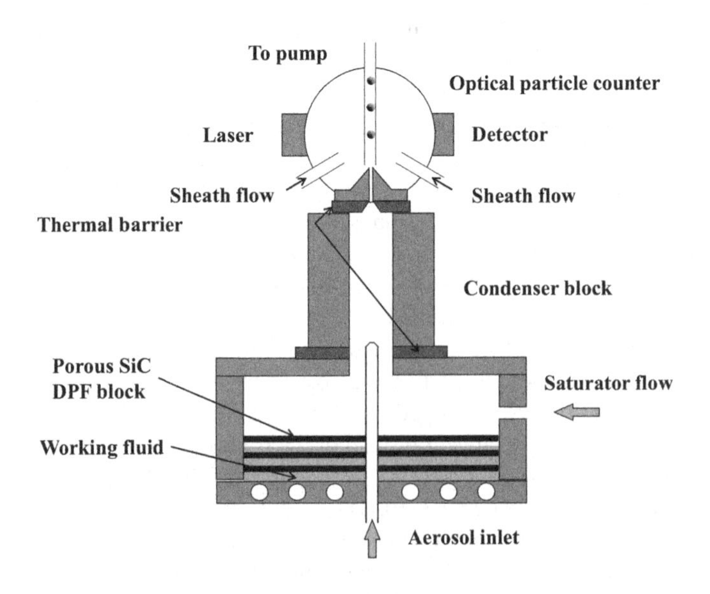

FIGURE 1.11 Schematic diagram of the HT-CPC.

2. Baltzer et al. developed a new CPC, as shown in Figure 1.12, and evaluated its counting efficiency [119]. In the new CPC, the condenser section is positioned above the saturator section, causing the working fluid to rise in a spiral shape in the saturator section. Baltzer et al. conducted tests to determine the counting efficiency of the CPC and the instrument's ability to measure the minimum size of graphite particles. Baltzer et al. conducted tests by fixing the temperature of the saturator section and adjusting the temperature of the condenser section to create different supersaturation conditions for the working fluid. The results revealed that the lower limit of particle size measurement decreased as the temperature of the saturator section decreased. However, the lower limit of particle size measurement decreased slowly with the decrease in temperature of the condenser section and did not decrease further after reaching 5 nm.
3. In recent research, Chen et al. [120–122] have been investigating the direct counting technology of fine particles at high temperatures. They were inspired by the nonmonotonic change characteristic of surface tension with temperature in "variable temperature effect" fluid and proposed a new design criterion for working fluid. After conducting numerous experiments, they successfully developed a new working fluid with non-ideal "variable temperature effect". However, due to the strong non-ideality of the new working fluid, its evaporation and condensation mechanism remains unclear. Chen et al. proposed a new theory on the unsteady evaporation and staged condensation growth of non-ideal solutions. These studies have made significant theoretical advancements and successfully addressed the challenge of measuring temperatures across a broad range from a mechanistic

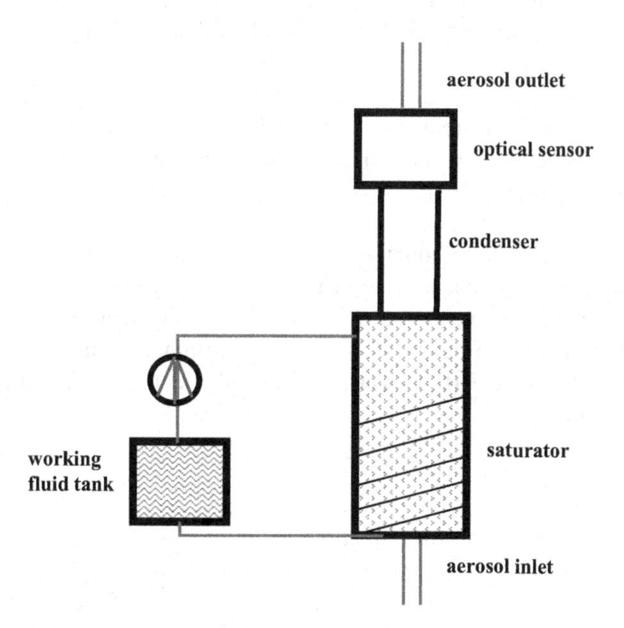

FIGURE 1.12 Schematic diagram of the new CPC invented by Baltzer et al.

FIGURE 1.13 Schematic diagram of wide-temperature CPC.

standpoint. This study addresses the challenge of accurately measuring the number concentration of fine particles while classifying them based on size. To tackle this issue, researchers have developed the wide-temperature CPC technology. The first wide-temperature CPC (Figure 1.13) capable of high-precision measurement of the number concentration of fine particles under the condition of particle size classification has been successfully invented and tested.

CPC is a popular method for counting fine particles and is used as a test method for measuring the number concentration of particles as specified by EU motor vehicle emission regulations. However, current research and application of CPC are primarily conducted at room temperature and only a limited number of experimental devices are capable of directly measuring the number concentration of fine particles at high temperatures.

REFERENCES

[1] Zhu T., Xu D.Y., Yu Y. 2013. *Atmospheric Particulate Control*. Beijing: Chemical Industry Press.

[2] Cha Y., He J.L., Cheng F., Li Y. 2019. *Atmospheric Aerosol Inversion and Remote Sensing for Pollution Environment Monitoring*. Beijing: Science Press.

[3] Zhan D.S., Kwan M.P., Zhang W.Z., et al. 2018. The driving factors of air quality index in China. *Journal of Cleaner Production* 197:1342–1351.
[4] Jiang G.B., Wang C.X., Zhang A.Q. 2020. *Toxicology and Health Effects of Atmospheric Fine Particulate Matters*. Beijing: Science Press.
[5] Zhang W. 2019. *Numerical Model of Capturing Fine Particulates*. Beijing: Chemical Industry Press.
[6] Zhou Y. 2012. *Characteristics and Sources of Aerosol in North China Plain and Smog Chamber Simulation*. Jinan: Shandong University.
[7] Lehtipalo K., Mikkila J., Vanhanen J. 2011. Particle size magnifier for nano-CN detection. *Aerosol Science and Technology* 45: 533–542.
[8] Collins A.M., Dick W.D., Romay F.J. 2013. A new coincidence correction method for condensation particle counters. *Aerosol Science and Technology* 47:177–182.
[9] Enroth J., Kangasluoma J., Korhonen F., et al. 2018. On the time response determination of condensation particle counters. *Aerosol Science and Technology* 52:778–787.
[10] Terres A., Giechaskiel B., Nowak A., et al. 2021. Calibration uncertainty of 23nm engine exhaust condensation particle counters with soot generators: A European automotive laboratory comparison. *Emission Control Science and Technology* 7:124–136.
[11] McMurry P.H. 2000. The history of condensation nucleus counters. *Aerosol Science and Technology* 33:297–322.
[12] Yu T.Z. 2018. *Research on Key Technologies of On-Line Measurement of Sampling and Number Concentration of Ultrafine Particles in Motor Vehicle*. Hefei: University of Science and Technology of China.
[13] Ministry of Environmental Protection of the People's Republic of China. *China Vehicle Emission Control Annual Report* (in Chinese). Beijing: Ministry of Environmental Protection of the People's Republic of China, 2022.
[14] Kittelson D.B., Arnold M., Watts W.F. 1999. *Review of Diesel Particulate Matter Sampling Methods*. Final Report. Minneapolis, MN: University of Minnesota, p. 63.
[15] Kittelson D.B. 1998. Engines and nanoparticles: A review. *Journal of Aerosol Science*. 29(5–6):48–86.
[16] Bond T.C., Bergstrom R.W. 2006. Light absorption by carbonaceous particles: An investigative review. *Aerosol Science and Technology* 40(1):27–67.
[17] Bond T.C. 2007. Can warming particles enter global climate discussions? *Environmental Research Letters* 2(4):045030–045038.
[18] Kärcher B. 2018. Formation and radiative forcing of contrail cirrus. *Nature Communications* 9(1):1–17.
[19] Zhang T. 2022. *Experimental Measurement of Mass and Quantity of Non-Volatile Particulate Matter from Aircraft Engines*. Beijing: Beihang University.
[20] Anderson B., Chen G., Blake D. 2006. Hydrocarbon emissions from a modern commercial airliner. *Atmospheric Environment* 40(19):3601–3612.
[21] Knighton W.B., Rogers T.M., Anderson B.E., et al. 2007. Quantification of aircraft engine hydrocarbon emissions using proton transfer reaction mass spectrometry. *Journal of Propulsion and Power* 23(5):949–949.
[22] Anderson B.E., Cofer W.R., Bagwell D.R., et al. 1998. Airborne observations of aircraft aerosol emissions I: Total nonvolatile particle emission indices. *Geophysical Research Letters* 25(10):1689–1692.
[23] Anderson B.E., Cofer W.R., Barrick J.D., et al. 1998. Airborne observations of aircraft aerosol emissions II: Factors controlling volatile particle production. *Geophysical Research Letters* 25(10):1693–1696.
[24] Liati A., Brem B.T., Durdina L., et al. 2014. Electron microscopic study of soot particulate matter emissions from aircraft turbine engines. *Environmental Science and Technology* 48(18):10975–10983.

[25] Li W., Shi Z., Yan C., et al. 2013. Individual metal-bearing particles in a regional haze caused by firecracker and firework emissions. *Science of the Total Environment* 443:464–469.
[26] Liu L., Kong S.F., Zhang Y.X., et al. 2017. Morphology, composition, and mixing state of primary particles from combustion sources - crop residue, wood, and solid waste. *Scientific Reports* 7(1):1–15.
[27] Li W.J., Liu L., Xu L., et al. 2020. Overview of primary biological aerosol particles from a Chinese boreal forest: Insight into morphology, size, and mixing state at microscopic scale. *Science of the Total Environment* 719(2020):137520.
[28] Li W.J., Shao L.Y., Zhang D.Z., et al. 2016. A review of single aerosol particle studies in the atmosphere of East Asia: Morphology, mixing state, source, and heterogeneous reactions. *Journal of Cleaner Production* 112(2016):1330–1349.
[29] Laskin A., Moffet R.C., Gilles M.K., et al. 2012. Tropospheric chemistry of internally mixed sea salt and organic particles: Surprising reactivity of NaCl with weak organic acids. *Journal of Geophysical Research: Atmospheres* 117(D15):D15302.
[30] Li W., Shao L., Shen R., et al. 2011. Internally mixed sea salt, soot, and sulfates at Macao, a coastal city in South China. *Journal of the Air & Waste Management Association* 61(11):1166–1173.
[31] Geng H., Ryu J.Y., Maskey S., et al. 2011. Characterisation of individual aerosol particles collected during a haze episode in Incheon, Korea using the quantitative ED-EPMA technique. *Atmospheric Chemistry and Physics* 11(3):1327–1337.
[32] Li W.J., Shao L.Y., Shen R., et al. 2010. Size, composition, and mixing state of individual aerosol particles in a South China coastal city. *Journal of Environmental Sciences* 22(4):561–569.
[33] Zhang D.Z., Iwasaka Y. 2004. Size change of Asian dust particles caused by sea salt interaction: measurements in southwestern Japan. *Geophysical Research Letters* 31(15):L15102.
[34] Lee D.S., Fahey D.W., Forster P.M., et al. 2009. Aviation and global climate change in the 21st century. *Atmospheric Environment* 43(22–23):3520–3537.
[35] Bond T.C., Doherty S.J., Fahey D.W., et al. 2013. Bounding the role of black carbon in the climate system: A scientific assessment. *Journal of Geophysical Research: Atmospheres* 118(11):5380–5552.
[36] Suo D.F., Zeng S.W. 2019. Research on the harm of air fine particulate matter PM2.5 to various human systems. *Medical Information* 32(18):32–34.
[37] Guo L., Johnson G.R., Hofmann W., et al. 2020. Deposition of ambient ultrafine particles in the respiratory tract of children: A novel experimental method and its application. *Journal of Aerosol Science* 139:105465.
[38] Hei H. 2013. *Perspectives 3: Understanding the Health Effects of Ambient Ultrafine Particles (HEI Review Panel on Ultrafine Particles)*. Boston, MA: Health Effects Institute.
[39] Schraufnagel E. 2020. The health effects of ultrafine particles. *Experimental and Molecular Medicine* 52:311–317.
[40] Ohlwein S., Kappeler R., Joss M.K., et al. 2019. Health effects of ultrafine particles: A systematic literature review update of epidemiological evidence. *International Journal of Public Health* 64:547–559.
[41] Tian Q., Li M., Montgomery S., et al. 2020. Short-term associations of fine particulate matter and synoptic weather types with cardiovascular mortality: An ecological time-series study in Shanghai, China. *International Journal of Environmental Research and Public Health* 17(3):E1111.
[42] Salim S.Y., Kaplan G.G., Madsen K.L. 2014. Air pollution effects on the gut microbiota: A link between exposure and inflammatory disease. *Gut Microbes* 5(2):215–219.

[43] Wang C., Feng L., Chen K. 2019. The impact of ambient particulate matter on hospital outpatient visits for respiratory and circulatory system disease in an urban Chinese population. *Science of the Total Environment* 666:672–679.
[44] Li R.Y., Yang L.Q., Jiang N., et al. 2020. Activated macrophages are crucial during acute PM2.5 exposure-induced angiogenesis in lung cancer. *Oncology Letters* 19(1):725–734.
[45] Lao X.Q., Zhang Z, Lau A.K., et al. 2018. Exposure to ambient fine particulate matter and semen quality in Taiwan. *Occupational and Environmental Medicine* 75(2):148–154.
[46] Ding, A., Yang, Y., Zhao, Z., et al. 2017. Indoor PM2.5 exposure affects skin aging manifestation in a Chinese population. *Scientific Reports* 7(1):1–7.
[47] World Health Organization (WHO). 2013. *Health Effects of Particulate Matter: Policy Implications for Countries in Eastern Europe, Caucasus and Central Asia.* Copenhagen: World Health Organization.
[48] Zhang W.X., Xiao C.L. 2021. Effects of PM2.5 on respiratory system and its mechanism. *Chinese Journal of Microecology* 33(11):1340–1349.
[49] See S.W., Karthikeyan S., Balasubramanian R. 2006. Health risk assessment of occupational exposure to particulate-phase polycyclic aromatic hydrocarbons associated with Chinese, Malay and Indian cooking. *Journal of Environmental Monitoring* 8(3):369–376.
[50] Emily U. 2017. The polluted brain. *Science* 355(6323):342–345. doi: 10.1126/science.355.6323.342.
[51] Li S., Zhang Y., Huang H. 2022. Black phosphorus-based heterostructures for photocatalysis and photoelectrochemical water splitting. *Journal of Energy Chemistry* 67:745–779.
[52] BéruBé K., Balharry D., Sexton K., et al. 2007. Combustion - derived nanoparticles: Mechanisms of pulmonary toxicity. *Clinical and Experimental Pharmacology and Physiology* 34(10):1044–1050.
[53] Office J.E., Chen J., Dan H., et al. 2021. New innovations in pavement materials and engineering: A review on pavement engineering research 2021. *Journal of Traffic and Transportation Engineering (English Edition)* 8(6):815–999.
[54] Jambers W., De Bock L., Van Grieken R. 1995. Recent advances in the analysis of individual environmental particles. A review. *Analyst* 120(3):681–692.
[55] Sun J., Zhou Z., Huang J., Li G. 2020. A bibliometric analysis of the impacts of air pollution on children. *International Journal of Environmental Research and Public Health* 17(4):1277.
[56] Dubey A., Lobo C.L., Ravi G.S., et al. 2022. Exosomes: Emerging implementation of nanotechnology for detecting and managing novel corona virus-SARS-CoV-2. *Asian Journal of Pharmaceutical Sciences* 17(1):20–34.
[57] Kim Y.T. 2019. *Framing Air Pollution.* Shanghai: Shanghai Jiao Tong University.
[58] Belkacem I., Helali A., Khardi S., et al. 2022. Road traffic nanoparticle characteristics: Sustainable environment and mobility. *Geoscience Frontiers* 13(1):101196.
[59] Maharjan L., Kang S., Tripathee L., et al. 2022. Atmospheric particle-bound polycyclic aromatic compounds over two distinct sites in Pakistan: Characteristics, sources and health risk assessment. *Journal of Environmental Sciences* 112:1–15.
[60] Pui D.Y., Chen S.C., Zuo Z. 2014. PM2. 5 in China: Measurements, sources, visibility and health effects, and mitigation. *Particuology* 13:1–26.
[61] Nandasena S., Wickremasinghe A.R., Sathiakumar N. 2013. Indoor air pollution and respiratory health of children in the developing world. *World Journal of Clinical Pediatrics* 2(2):6.
[62] Jody N.N. 2019. *Research on Mechanism and Modes of Green Supply Chain.* Wuhan: Wuhan University of Technology.
[63] Jendryke M.I. 2016. *Shanghai's Urban Vibrancy Using Microwave Remote Sensing and Big Social Sensing Data.* Wuhan: Wuhan University.

[64] Zhang J., Tong L., Peng C., et al. 2019. Temporal variability of visibility and its parameterizations in Ningbo, China. *Journal of Environmental Sciences* 77:372–382.

[65] Wang G., Shi X., Cui H., Jiao J. 2020. Impacts of migration on urban environmental pollutant emissions in China: A comparative perspective. *Chinese Geographical Science* 30:45–58.

[66] Zhao Y., Zhang X., Chen M., Gao S., Li R. 2022. Regional variation of urban air quality in China and its dominant factors. *Journal of Geographical Sciences* 32(5):853–872.

[67] Liu Y. 2011. *Study on the Characteristics of Particulate Matter Pollution in Urban Traffic Environment Based on Traffic Flow Control*. Beijing: Tsinghua University.

[68] Zhu X.J. 2018. *Calibration Study of Automatic Monitoring Instrument for Air Particulate Matter by β-Ray Method*. Tianjin: Tianjin University of Technology.

[69] Guo L., Shao P.W., Ma Z.W., et al. 2021. Study on the applicability of light scattering method and β ray method particulate matter measuring instrument. *China Metrology* 2021(2):81–85.

[70] Chen F., Chen F., Yu H., et al. 2017. B-ray method versus gravimetric monitoring of ambient air PM_ (2.5). *Environmental Engineering* 35(8):146–151.

[71] Xia Q. 2009. *Study on the Impact of Aircraft Engine Emissions on the Atmospheric Environmental Impact of Airports*. Nanjing: Nanjing University of Aeronautics and Astronautics.

[72] Wang T., Gao T., Zhang H., et al. 2019. Review of Chinese atmospheric science research over the past 70 years: Atmospheric physics and atmospheric environment. *Science China Earth Sciences* 62:1903–1945.

[73] Peng J., Tang F., Zhou R., et al. 2016. New techniques of on-line biological sample processing and their application in the field of biopharmaceutical analysis. *Acta Pharmaceutica Sinica B* 6(6):540–551.

[74] Editorial Board of China Journal of Highway and Transport. 2017. A review of academic research on automotive engineering in China 2017. *China Journal of Highway and Transport* 30(6):1–197.

[75] https://gawsis.meteoswiss.ch/GAWSIS/#/

[76] Hyslop N.P., White W.H. 2008. An evaluation of interagency monitoring of protected visual environments (IMPROVE) collocated precision and uncertainty estimates. *Atmospheric Environment* 42:2691–2705.

[77] Rizzo M.J., Scheff P.A. 2007. Fine particulate source apportionment using data from the USEPA speciation trends network in Chicago, Illinois: Comparison of two source apportionment models. *Atmospheric Environment* 41:6276–6288.

[78] https://emep.int/publ/reports/2022/EMEP_Status_Report_1_2022.pdf

[79] https://data.ec.gc.ca/data/air/monitor/networks-and-studies/canadian-air-and-precipitation-monitoring-network-capmon/?lang=en

[80] Xin J., Wang Y., Pan Y., et al. 2015. The campaign on atmospheric aerosol research network of China: CARE-China. *Bulletin of the American Meteorological Society* 96:1137–1155.

[81] https://aeronet.gsfc.nasa.gov/

[82] Stefano V., Mario C., Luca B., et al. 2022. Morphology and electronic properties of incipient soot by scanning tunneling microscopy and spectroscopy. *Combustion and Flame* 243:111980.

[83] Martin J.W., Salamanca M., Kraft M. 2022. Soot inception: Carbonaceous nanoparticle formation in flames. *Progress in Energy and Combustion Science* 88:100956.

[84] Chen L., Hu X., Wang J., et al. 2019. Impacts of alternative fuels on morphological and nanostructural characteristics of soot emissions from an aviation piston engine. *Environmental Science & Technology* 53:4667–74.

[85] Commodo M., Kaiser K., De Falco G., et al. 2019. On the early stages of soot formation: Molecular structure elucidation by high-resolution atomic force microscopy. *Combustion and Flame* 205:154–164.
[86] Noskov A., Ervik T. K., Tsivilskiy I., et al. 2020. Characterization of ultrafine particles emitted during laser-based additive manufacturing of metal parts. *Scientific Reports* 10:20989.
[87] Marhabaa I., Ferrya D., Laffon C., et al. 2019. Aircraft and MiniCAST soot at the nanoscale. *Combustion and Flame* 204:278–289.
[88] Liu C., Chen G., Jin Y., et al. 2022. Experimental studies on soot in MILD oxy-coal combustion flame: Sampling, microscopic characteristics, chemical compositions and formation mechanism. *Combustion and Flame* 241:112094
[89] Giechaskiel B., Maricq M., Ntziachristos L., et al. 2014. Review of motor vehicle particulate emissions sampling and measurement: From smoke and filter mass to particle number. *Journal of Aerosol Science* 67:48–86.
[90] Kangasluoma J., Attoui M. 2019. Review of sub-3 nm condensation particle counters, calibrations, and cluster generation methods. *Aerosol Science and Technology* 53:1277–1310.
[91] Keller A., Tritscher T., Burtscher H. 2013. Performance of water-based CPC 3788 for particles from a propane-flame soot-generator operated with rich fuel/air mixtures. *Journal of Aerosol Science* 60:67–72.
[92] Hermann M., Wiedensohler A. 2001. Counting efficiency of condensation particle counters at low-pressures with illustrative data from the upper troposphere. *Journal of Aerosol Science* 32:975–991.
[93] Sem G.J. 2002. Design and performance characteristics of three continuous-flow condensation particle counters: A summary. *Atmospheric Research* 62:267–294.
[94] Liu W., Kaufman S.L., Osmondson B.L., et al. 2006. Water-based condensation particle counters for environmental monitoring of ultrafine particles. *Journal of the Air & Waste Management Association* 56:444–455.
[95] Liu W., Osmondson B.L., Bischof O.F., et al. 2005. *Calibration of Condensation Particle Counters.* SAE Technical Paper Series 2005-01-0189.
[96] Hering S.V., Stolzenburg M.R. 2005. A method for particle size amplification by water condensation in a laminar, thermally diffusive flow. *Aerosol Science and Technology* 39:428–436.
[97] Hering S.V., Stolzenburg M.R., Quant F.R., et al. 2005. A laminar-flow, water-based condensation particle counter (WCPC). *Aerosol Science and Technology* 39:659–672.
[98] Feldpausch P., Fiebig M., Fritzsche L., et al. 2006. Measurement of ultrafine aerosol size distributions by a combination of diffusion screen separators and condensation particle counters. *Journal of Aerosol Science* 37:577–597.
[99] Bau S., Toussaint A., Payet R., et al. 2017. Performance study of various Condensation Particle Counters (CPCs): Development of a methodology based on steady-state airborne DEHS particles and application to a series of handheld and stationary CPCs. *Journal of Physics: Conference Series* 838:012002.
[100] Wimmer D., Lehtipalo K., Nieminen T., et al. 2014. Technical note: Using DEG CPCs at upper tropospheric temperatures. *Atmospheric Chemistry & Physics Discussions* 14:12797–12817.
[101] Franklin L.M., Bika A.S., Watts W.F., et al. 2010. Comparison of water and butanol based CPCs for examining diesel combustion aerosols. *Aerosol Science and Technology* 44:629–638.
[102] Kulmala M., Mordas G., Petäjä T., et al. 2007. The condensation particle counter battery, CPCB: A new tool to investigate the activation properties of nanoparticles. *Journal of Aerosol Science* 38:289–304.

[103] Kim S., Iida K., Kuromiya Y., et al. 2014. Effect of nucleation temperature on detecting molecular ions and charged nanoparticles with a diethylene glycol-based particle size magnifier. *Aerosol Science and Technology* 49:35–44.

[104] Sgro L., Juan F.D.L.M. 2004. A simple turbulent mixing CNC for charged particle detection down to 1.2 nm. *Aerosol Science and Technology* 38(1):1–11.

[105] Kangasluoma J., Ahonen L., Attoui M., et al. 2015. Sub-3 nm particle detection with commercial TSI 3772 and airmodus A20 fine condensation particle counters. *Aerosol Science and Technology* 49:674–681.

[106] Giechaskiel B., Wang X., Gilliland D., et al. 2011. The effect of particle chemical composition on the activation probability in n-butanol condensation particle counters. *Journal of Aerosol Science* 42:20–37.

[107] Zhang X., Chen L., Liang Z., et al. 2015. The theoretical study and numerical simulation of the condensed nuclear particle counter. *Science and Technology Guide* 33:73–78.

[108] Yan F., Hu H., Yu T. 2004. The particle mass concentration and visibility were measured with an optical particle counter. *Journal of Quantum Electronics* 21:98–102.

[109] Yu T., Zhang J., Wang J., et al. A number and concentration measuring device of atmospheric ultrafine particulate matter. Chinese patent: 201410578772, 2015-01-21.

[110] Jiang J., Chen M., Kuang C., et al. 2011. Electrical mobility spectrometer using a diethylene glycol condensation particle counter for measurement of aerosol size distributions down to 1 nm. *Aerosol Science and Technology* 45:510–521.

[111] Jiang J., Zhao J., Chen M., et al. 2011. First measurements of neutral atmospheric cluster and 1-2 nm particle number size distributions during nucleation events. *Aerosol Science and Technology* 45:ii–v.

[112] Lam D.H., Jangkamolkulchai A., Luks K.D. 1990. Liquid-liquid-vapor phase equilibrium behavior of certain binary carbon dioxide+ n-alkanol mixtures. *Fluid Phase Equilibria* 60:131–141.

[113] Chen C.C., Tsai W.T. 2002. Condensation of supersaturated n-butanol vapor on charged/neutral nanoparticles of D-mannose and L-rhamnose. *Journal of Colloid & Interface Science* 246:270–280.

[114] Shmyt'Ko I.M., Jiménez-Riobóo R.J., Hassaine M., et al. 2010. Structural and thermodynamic studies of n-butanol. *Journal of Physics: Condensed Matter* 22:195102.

[115] Gharibeh M., Kim Y., Dieregsweiler U., et al. 2005. Homogeneous nucleation of n-propanol, n-butanol, and n-pentanol in a supersonic nozzle. *Journal of Chemical Physics* 122:094512.

[116] Zheng Z., Johnson K.C., Liu Z., et al. 2011. Investigation of solid particle number measurement: Existence and nature of sub-23nm particles under PMP methodology. *Journal of Aerosol Science* 42:883–897.

[117] Filippo A.D., Maricq M.M. 2008. Diesel nucleation mode particles: Semivolatile or solid. *Environmental Science & Technology* 42:7957–7962.

[118] Collings N., Rongchai K., Symonds, J.P.R., et al. 2014. A condensation particle counter insensitive to volatile particles. *Journal of Aerosol Science* 73:27–38.

[119] Baltzer S., Onel S., Weiss M., et al. 2014. Counting efficiency measurements for a new condensation particle counter. *Journal of Aerosol Science* 70:11–14.

[120] Chen L., Ma Y., Guo Y., et al. 2017. Quantifying the effects of operational parameters on the counting efficiency of a condensation particle counter using response surface Design of Experiments (DoE). *Journal of Aerosol Science* 106:11–23.

[121] Zhang X., Chen L., Liang Z., et al. 2015. Theoretical research and numerical simulation of compact core particle counter. *Technology Guide* 6:73–78.

[122] Chen L., Ma Y., Zhang X., et al. 2017. Factor sensitivity analysis and empirical formula construction of the cutting particle size. *Applied Technology* 5:22–29.

2 Theory and Mechanisms of the Fine Particles Condensation Growth

Guangze Li, Shenghui Zhong, and Longfei Chen

2.1 INTRODUCTION

The condensation of water vapor can be classified into two types: homogeneous condensation and heterogeneous condensation. When the water vapor in the environment becomes supersaturated, it can condense. In a pure water vapor environment, the condensable water vapor molecules will undergo autogenous polymerization. This process is referred to as homogeneous condensation [1]. When particles are present in the environment, water vapor molecules will utilize these particles as condensation nuclei. This process is known as heterogeneous condensation. Homogeneous condensation in a supersaturated environment requires a higher degree of supersaturation, and the critical degree of supersaturation is 2–5 [2]. In a supersaturated environment, heterogeneous condensation requires a lower degree of supersaturation, with the critical degree of supersaturation being low (about 1–2) [3,4].

The heterogeneous condensation of water vapor on the surface of fine particles can be divided into two stages: nucleation and condensation growth. Nucleation refers to the process from the start of water vapor condensation to the formation of a critical embryo on the particle's surface. Condensation growth is the continuous condensation of water vapor on the critical embryo's surface, leading to an increase in particle size until the condensation rate of water vapor reaches equilibrium with the evaporation rate of droplets on the particle surface [5]. The prerequisite for the growth of water vapor on the surface of fine particles is nucleation on the particle's surface. Therefore, the study of nucleation theory holds significant importance. Many researchers have studied the heterogeneous nucleation of water vapor on particle surfaces, and extensive experimental studies on the nucleation and condensation of water vapor on fine particle surfaces have also been carried out [6,7]. These research findings indicate that the nucleation of water vapor on particle surfaces is primarily affected by particle size, supersaturation, particle surface properties, and particle chemical composition.

To date, research on the nucleation theory of water vapor on particle surfaces has mainly focused on classical heterogeneous nucleation theory [8], adsorption theory [8], density functional theory [9], and molecular dynamics Monte Carlo theory [10]. Heterogeneous nucleation of water vapor on particle surfaces is a prevalent natural phenomenon. For example, water vapor in the atmosphere condenses on cloud

DOI: 10.1201/9781003423195-2

condensation nuclei's surfaces, leading to the formation of clouds and fog [11,12]. Classical heterogeneous nucleation theory relies on the capillary approximation to solve the thermodynamic process [13]. While this theory greatly simplifies the thermodynamic process. Consequently, this simplification is considered a main reason for the discrepancy between theoretical and experimental nucleation rates. The major flaw in this theory lies in applying macroscopic surface tension to the microscopic thermodynamic process of the particle-water phase. Nevertheless, despite its drawbacks, this theory remains the most practical approach currently available. The application of heterogeneous nucleation theory has encountered substantial difficulties when dealing with fine particles with complex surface structure and chemical compositions. In adsorption theory, to avoid the use of uncertain kinetic constants, vapor nucleation and condensation are treated as adsorption processes.

In 1996, Talanquer and Oxtoby [9] proposed a density functional theory for the study of nucleation. This theory proposes the use of a density function, denoted as $\rho(r)$, to describe the interaction between the newly formed phase consisting of water and particles and the initial phase within a specific range of widths. Unlike classical heterogeneous nucleation theory, this approach departs from the use of the capillary approximation. Despite the increasing accuracy of various methods for measuring nucleation rates, there remains a lack of a direct experimental technique capable of precisely measuring nucleation rates. The available methods can only provide insights into the macroscopic structure of critical nuclei, not the observed microstructure. Molecular dynamics (MD) simulation, on the other hand, offers the ability to depict intricate details of the nucleation process at the microscopic level. Therefore, MD simulation is a valuable method for studying nucleation, although it comes with the drawback of requiring significant computational resources due to its high computational demands.

Several studies have investigated particle growth characteristics, although measuring a droplet with a particle at its core after growth presents challenges. Research shows that particle growth after nucleation is mainly affected by particle wettability, particle number concentration, particle residence time, and water vapor supersaturation [5]. Researchers have examined the growth characteristics of combustion-derived fine particles and mineral particles. These investigations offer valuable insights into specific industrial operations. Considering the effect of particle characteristics, comparing the growth characteristics of different particles with varying wettability may not accurately assess the impact of wettability on the condensation of fine particles. The exploration of vapor phase transitions that facilitate the condensation-driven growth of fine particles is still in its experimental stage. Significant progress is required before this knowledge can be effectively applied within industrial settings.

The condensation growth of fine particles is a typical heat and mass transfer process. Based on particle size, condensation growth can be categorized into three fundamental mechanisms: (1) continuous mechanism, (2) free molecular mechanism, and (3) Knudsen mechanism. When particles surpass a critical size under specific saturation and temperature conditions, it is generally accepted that their diameter exceeds the average free path of vapor molecules. In this scenario, particle growth rate depends on the interchange of molecules and heat between particles and the surrounding media, essentially relying on diffusion processes. Conversely, when particle size is significantly smaller than the surrounding vapor's average free path, particle condensation rate can be calculated using kinetic principles [14]. For particle sizes falling between the two previously mentioned cases, a Knudsen transport mechanism is adopted to explain the particle growth [15]. To model the Knudsen transport mechanism, researchers often assume that mass transport on the droplet surface is diffusion-controlled beyond a specific distance. Between the droplet surface and this designated point, aerodynamics dominates. At present, researchers have developed a large number of mass transport equations based on Knudsen transport mechanism, and the results are in good agreement with existing experimental findings.

Due to the small size of fine particles, which often fall beyond the range of existing optical detection methods, the condensation growth technique is commonly employed. This method involves coalescing and enlarging fine particles to a size that can be optically detected, followed by particle counting through the capture of optical signals. Research into the detection of previously invisible particles using particle condensation growth dates back to the late 19th century [16–18]. This method entails subjecting particles to heterogeneous condensation within a saturated vapor environment. Condensation theory highlights the strong dependence of heterogeneous condensation on vapor saturation (s) and temperature (T). Consequently, achieving condensation growth of fine particles primarily revolves around adjusting these two key physical parameters to attain a supersaturated state of vapor. This state enables swift vapor condensation and growth on the surfaces of the fine particles.

The condensation growth of fine particles mainly occurs in the state of vapor supersaturation. This process can be categorized into three types based on different methods of achieving a supersaturated state: (1) adiabatic expansion condensation growth, (2) turbulent mixing condensation growth, and (3) continuous laminar flow non-isothermal diffusion condensation growth. In adiabatic expansion condensation growth, the process involves first elevating the humidity of the aerosol system at room temperature until water vapor saturation is achieved. Subsequently, the aerosol is rapidly cooled within an expansion chamber through volume expansion or pressure release [19]. This cooling results in vapor supersaturation at a lower temperature, causing condensation to occur on the fine particles. The expansion rate dictates the rate of steam supersaturation, requiring a sufficiently rapid expansion to minimize heat conduction within the system. Turbulent mixing condensation growth achieves swift and nearly adiabatic mixing by combining a cold aerosol flow with a considerable amount of hot and humid gas. High-intensity turbulence is generated within the mixing region through impinging jet [20]. Mixing steam-saturated warm air with cooler aerosol flow creates supersaturation in the aerosol without using an external cooler. This method ensures rapid component uniformity and minimal particle loss within only 50 ms. The mixed gas enters a low-temperature condenser to achieve further condensation growth. However, due to the turbulent nature of the gas flow in the mixing chamber, precise control over temperature and steam pressure distribution is challenging. Continuous laminar heat conduction condensation growth consists of a saturator and a condenser. Aerosol passes through a saturator equipped with a liquid storage tank, where, at the set temperature, it is saturated by the working fluid. The saturated aerosol then proceeds to the condenser, which maintains a low temperature. In the condenser, gas cooling occurs through conduction and convection, leading to vapor supersaturation in the cooled aerosol. The gas flow in both the saturation and condensation chambers follows a laminar structure with a parabolic gas flow velocity profile. The continuous laminar heat conduction condensation growth method is widely employed due to its straightforward setup and manageable process.

To enhance control over the condensation growth process, recent efforts have introduced saturated aerosol fluid at the pipeline's wall surface. This approach ensures uniform temperature and supersaturation throughout the central area of the pipeline. The subsequent sections of this chapter will delve into the fundamental principles, technical implementation, and control strategies of fine particle condensation growth. Additionally, recent advancements and future prospects within this field will be summarized.

2.2 NUCLEATION THEORY

2.2.1 Kelvin Effect

Tiny hygroscopic particles will condense and form droplets in a high-humidity environment. It is a common practice to generate the conditions necessary for condensation and growth of tiny particles by establishing a supersaturation field within the condensation section.

In the vessel, liquid evaporation and condensation occur simultaneously. When the two processes reach equilibrium, the corresponding vapor pressure equals its saturated vapor pressure, expressed as p_s [21]. Different liquids possess distinct saturation vapor pressures at varying temperatures. For most types of compounds, the saturated vapor pressure can be calculated using Antoine's equation (2.1):

$$\log_{10} p_s = a - \frac{b}{T - c} \tag{2.1}$$

The values of the coefficients a, b, and c in the above formula depend on the chemical composition of the substance. For some compounds, some modifications of the Antoine equation are necessary; see equation (2.2):

$$\log_{10} p_s = -\frac{52.3b}{T} + c \tag{2.2}$$

The above formula calculates the saturated vapor pressure of the liquid on the plane, and the corresponding saturation value can be calculated according to equation (2.3):

$$S = \frac{p}{p_{\text{sat}}(T)} \tag{2.3}$$

Where S is the supersaturation degree, and saturation measures the disparity in the rates of evaporation and vapor molecule condensation in a liquid. In order for a droplet to condense on a flat surface, its partial vapor pressure must be greater than the saturated vapor pressure of the liquid; that is, saturation must be greater than 1.

However, for particles, the liquid molecules on the particle surface are looser than those on the flat surface due to the presence of curvature. The trapping effect of the particle on the liquid molecules is less than that on the flat surface, so the liquid molecules on the particle surface are more likely to escape. Therefore, the liquid on the surface of the particle has a higher saturated vapor pressure. The smaller the particle size, the greater the curvature and the higher the saturated vapor pressure. The phenomenon, where curvature affects the vapor pressure of a liquid, is known as the Kelvin effect [22]. Correspondingly, the curvature of the surface where the pore structure exists is negative, and the vapor pressure of the liquid in the pores is lower than that on the flat surface.

For particles, the condition for activated growth is that the rate of adsorption or condensation of working fluid vapor molecules on their surface is greater than the rate of working fluid vapor molecules escaping from the particle surface. This process depends mainly on the particle size, the surface condensation rate, the working fluid properties on the surface, and the degree of supersaturation in the environment. When a group of aerosol particles is exposed to a supersaturated field filled with working fluid vapor, there must be a critical state at which the particles are in a state of condensation and growth and cannot be activated. The Kelvin equation describes this critical state and expresses the relationship between the activation conditions of the particles and the environmental parameters. See equation (2.4) [23,24] for the Kelvin equation:

$$d_k = \frac{4\gamma M}{\rho RT \ln S} \tag{2.4}$$

In the above formula, γ is the surface tension of the working fluid, M is the molar molecular mass of the working fluid, ρ is the density, R is the gas constant, T is the surface temperature of the droplet, and d_k is the critical diameter at which the particle can be activated in this environmental state, that is, the Kelvin equivalent diameter (Kelvin equivalent diameter), which means that when the particles are suspended in a certain working fluid steam, and the ambient temperature of the particles is T, and the supersaturation degree is S, only when the particle size is larger than the particle size d_k can be activated and grown. In terms of Gibbs free energy, the difference ΔG between the positive and negative Gibbs free energy represents whether the process can proceed spontaneously. For the condensation of droplets, the Gibbs free energy formula is shown in the following equation:

$$\Delta G = \left(\frac{4\pi R_p^3}{3v_l}\right)(g_l - g_v) + 4\pi R^2 \sigma \tag{2.5}$$

Where ΔG is the Gibbs free energy of the vapor and liquid molecules and is the surface tension. According to the formula, the relationship between and can be drawn in Figure 2.1:

As can be seen from the above figure, the maximum ΔG Gibbs free energy corresponds to a droplet radius value. R_p^* is quickly activated.

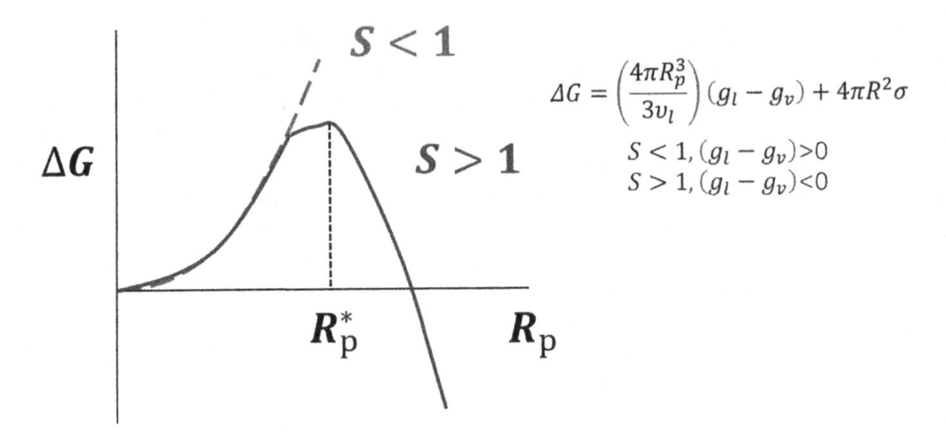

FIGURE 2.1 The relationship between Gibbs free energy and droplet radius.

The Kelvin Nucleation Theory is one of the theories often used by researchers to study particle nucleation, but it has its limitations. This theory is only suitable for predicting the growth of wet, insoluble particles. For soluble particles, the dissolving effect of the working fluid on the particles reduces the liquid vapor pressure of the droplets, resulting in a degree of supersaturation required for soluble particle growth that is lower than predicted by Kelvin's theory. However, Ahn and Liu in 1990 [25] performed experiments on the counting efficiency of n-butanol condensation particle counter (CPC) on NaCl and dioctyl titanate (DOP) particles. The saturation is less than that predicted by Kelvin theory, and the supersaturation required for condensation and growth of DOP particles is in agreement with that predicted by Kelvin theory. Kelvin's theory prediction model is applicable and reliable for insoluble particles [25–27].

In the design of CPC, the particle size of the original particles is fixed, and the best way for designers to reduce d_k is to increase the degree of supersaturation. However, excessive supersaturation will cause homogeneous nucleation, which will be introduced in the following sections. Therefore, when designing a CPC, it is essential to establish a reasonable range for the level of supersaturation.

2.2.2 Homogeneous Nucleation

In the previous section, it was introduced that particles can be activated and enlarged within a supersaturated working medium steam. In the process, the initial particles act as condensation nuclei and can also condense spontaneously in the absence of preexisting condensation nuclei in the working medium steam. The phenomenon of generating particles within a homogeneous phase is known as homogeneous nucleation. The classical theory of homogeneous nucleation describes the formation principle of particles in a homogeneous vapor phase. Homogeneous nucleation can be controlled by the degree of supersaturation. In a supersaturated vapor, molecules will form molecular chains between a small number of molecules to enter the second phase. However, these clusters lack thermodynamic stability and cannot endure over extended periods. As the degree of supersaturation rises, the number of clusters will also increase. Some of these clusters will grow to the critical size of nuclei that can activate growth, and the clusters exceeding this critical size will have a probability of condensation growth. Conversely, clusters smaller than the critical size will evaporate due to instability [28]. Its critical dimension can also be determined according to the Kelvin equation (2.4).

In homogeneous nucleation theory, the nucleation rate is defined as the net value of clusters formed per unit time with a size greater than the Kelvin critical diameter:

$$J = a\exp\left[-\frac{16\pi}{3}\times\frac{V^2\gamma^3}{(kT)^3(\ln S)^2}\right] \tag{2.6}$$

The above formula is obtained on the basis of thermodynamics and with some ideal assumptions; for example, the cluster is assumed to be a sphere. The influence of each parameter on the nucleation rate can be analyzed using the above formula, especially in the case of low supersaturation [29]. On the other hand, if the vapor pressure of the nucleating material is very low, the nucleation behavior is governed by chemical reactions [30,31]. In short, any monomer undergoing a chemical reaction in the gas phase can be regarded as a thermodynamically stable condensate nucleus that can nucleate, and the nucleation rate at that time is represented by the reaction rate.

2.2.3 Heterogeneous Nucleation

Homogeneous nucleation rarely occurs in various practices compared with heterogeneous nucleation and can be obtained from equation (2.7):

$$J \propto \exp\frac{\gamma^3}{T^3} \tag{2.7}$$

From equation (2.7), it can be deduced that reducing the surface tension at low temperature or low saturation can aid particle nucleation. When the droplet is on the container wall or on the particle, the droplet will form a cage-like structure due to the surface tension. This structure can minimize its holding energy, and the droplet will form a certain angle with the interface at equilibrium, as shown in Figure 2.2:

Where θ is called the wetting angle, and its formula is equation (2.8):

$$\cos\theta = (\gamma_{ML} - \gamma_{SM})/\gamma_{SL} \tag{2.8}$$

As shown in the figure above, when a droplet is placed on a solid surface, it behaves differently depending on its adhesion to the material surface. If this adhesion is attractive to the droplet, the droplet will be pulled toward the surface and spread out along the solid material, a behavior known as the hydrophilicity of the material to the droplet. Conversely, if the adhesive force is repulsive to the droplet, the droplet will reduce surface contact with the solid material, a behavior called hydrophobicity. The examples of hydrophilicity and hydrophobicity are shown in Figure 2.2. Combined with the definition of contact angle, it can be seen that the contact angle of the droplet

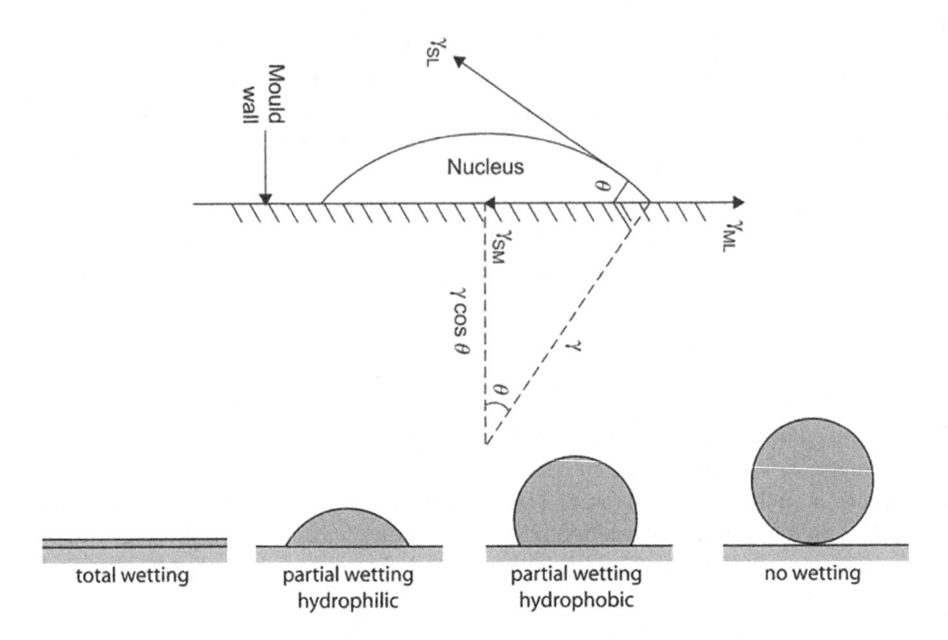

FIGURE 2.2 Schematic diagram of contact angle variation during heterogeneous condensation.

on the material surface is completely hydrophilic, which is a complete wetting; as the hydrophilicity increases, the contact angle also gradually decreases. In this case, it is difficult for the working fluid vapor to wrap around its surface, so only homogeneous nucleation can occur. Equation (2.5) describes the relationship between the Gibbs free energy and various parameters when the working fluid steam condenses on the particle surface. When it ΔG is at the extreme value, there is R_p:

$$\frac{d\Delta G}{dr} = 0 \tag{2.9}$$

$$R_p = \frac{2V\gamma}{KT \ln S} \tag{2.10}$$

$$\Delta G^*_{\text{homo}} = \frac{16\pi V^2\gamma^3}{3\left[KT\ln(S)\right]^2} = \frac{1}{3}\pi D_K^2\gamma \tag{2.11}$$

For heterogeneous nucleation, it needs to be multiplied by a $\Delta G^*_{\text{heter}}$ relevant factor f considering the contact angle between the working fluid vapor and the particle surface, and $0 \le f \le 1$ [8]:

$$\Delta G^*_{\text{heter}} = f\Delta G^*_{\text{homo}} \tag{2.12}$$

TABLE 2.1
Differences between Homogeneous and Heterogeneous Nucleation

	Homogeneous Nucleation	Heterogeneous Nucleation
Location	Occurs away from the particle surface	Occurs on the particle surface
Nucleation Site	Not involve	Involve
Rate	Slow	Quick
Free Energy	High energy barrier	Low energy barrier
Saturation	High saturation	Low supersaturation

$$f=\frac{1}{2}\times\left[\begin{array}{l}1+\left(\frac{1-X\times\cos\theta}{g}\right)^3+X^3\times(2-3\times\left(\frac{X-\cos\theta}{g}\right)+\left(\frac{X-\cos\theta}{g}\right)^3\\+3\times X^2\times\cos\theta\times\left(\frac{X-\cos\theta}{g}-1\right)\end{array}\right] \tag{2.13}$$

It can be seen that heterogeneous nucleation ΔG^*_{heter} is reduced by the presence of the contact angle compared to homogeneous nucleation, indicating that under normal circumstances, heterogeneous nucleation is more likely to occur.

The most critical difference between homogeneous and heterogeneous nucleation is that homogeneous nucleation occurs far from the particle surface, and its nucleation process does not involve nucleation sites and is slow, whereas heterogeneous nucleation occurs at the particle surface, and the process involves nucleation sites and is relatively fast. Table 2.1 shows the difference between the two processes:

2.3 CONDENSATION THEORY

In the transition stage of condensation, the liquid embryos formed on the particle surface absorb the surrounding water vapor and grow, but the particles are not completely enveloped. At this time, there are two mechanisms for the heterogeneous condensation and growth of liquid embryos on the particle surface: (1) the liquid embryos directly adsorb water molecules from the vapor and (2) water molecules adsorbed on the fine particle surface are transferred to the liquid embryo by diffusion [32]. Meanwhile, water molecules also evaporate outwards by these two mechanisms. Two mechanisms for the growth of water vapor on the surface of fine particles are shown in Figure 2.3. To simplify the model of the transition phase, we make the following assumptions: the transfer form of water molecules between the gas and liquid phases is a single water molecule; the interaction between the liquid embryos is ignored.

Assuming that there is a liquid embryo on the surface of the fine particle, which contains g water molecules, we define it as B_g. Then, the changes brought about by the liquid embryo gaining or losing a single water molecule can be expressed by the following formula:

$$B_g+B_1\rightleftharpoons B_{g+1},g=1,2,3\ldots \tag{2.14}$$

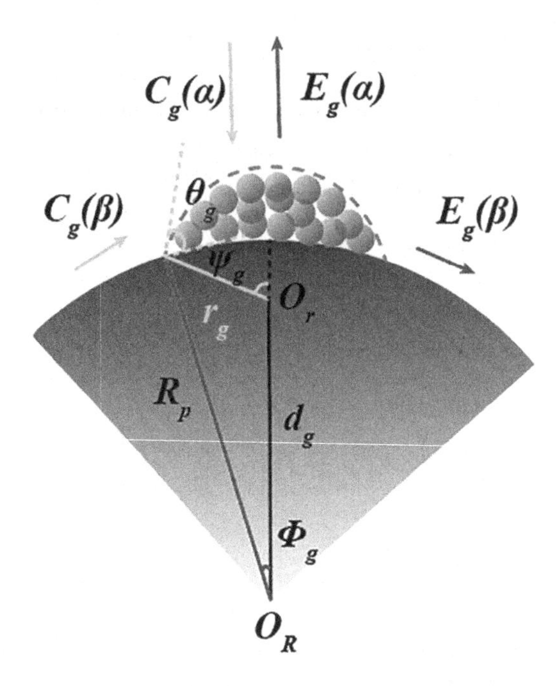

FIGURE 2.3 The growth mechanism of the droplet embryo on particle surface.

We define the condensation rates of liquid embryos to obtain water molecules through two mechanisms as $C_g(\alpha)$ and $C_g(\beta)$, respectively, and the evaporation rates of liquid embryos to lose water molecules through these two mechanisms are defined as $E_g(\alpha)$ and $Eg(\beta)$. Therefore, the condensation and evaporation rates of water molecules from the surface of the liquid embryo are, respectively:

$$C_g = C_g(\alpha) + C_g(\beta) \tag{2.15}$$

$$E_g = E_g(\alpha) + E_g(\beta) \tag{2.16}$$

Where α indicates that the liquid embryo directly adsorbs water molecules from the steam, β represents the transfer of surface-adsorbed water molecules to the liquid embryo by diffusion, R is the particle radius, and r_g is the liquid embryo radius. θ_g, ψ_g, and ϕ_g are the contact angle between the particle and the liquid embryo, the half of the angle between the particle sphere and the contact line between the particle and the liquid embryo, respectively. d_g is the center-to-center distance between the liquid embryo and the fine particles, as showed in Figure 2.4.

Defining f_g as the number of water molecules per unit surface area on a fine particle, we can get the rate of change of the number:

$$\frac{df_g}{dt} = C_{g-1}f_{g-1} - \left(C_g + E_g\right)f_g + E_{g+1}f_{g+1} \tag{2.17}$$

In heterogeneous condensation, the liquid embryo forms a "cap-shaped" droplet on the surface of the fine particle due to capillary action. With $V_1\,(rg)$, the volume of the

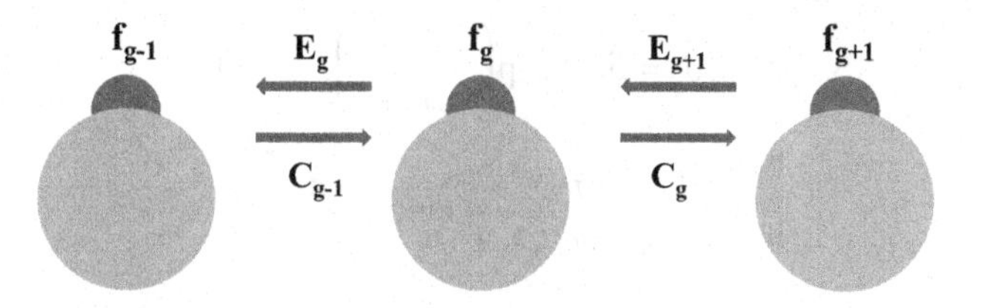

FIGURE 2.4 The variation of water molecules on particle surface.

liquid embryo with a radius of r_g. Where g represents the number of water molecules contained in the liquid embryo, V_L representing the volume of a water molecule, we can get:

$$V_l\left(r_g\right) = gV_L \tag{2.18}$$

From the geometric relationship, we know:

$$\cos\psi_g = \frac{mR - r_g}{d_g} \tag{2.19}$$

$$\cos\phi_g = \frac{R - mr_g}{d_g} \tag{2.20}$$

$$d_g = \sqrt{R^2 + r_g^2 - 2Rr_g m} \tag{2.21}$$

$$S_{vl}\left(r_g\right) = 2\pi r_g^2\left(1 - \cos\psi_g\right) \tag{2.22}$$

$$S_{ls}\left(r_g\right) = 2\pi r_g^2\left(1 - \cos\phi_g\right) \tag{2.23}$$

$$L_{vs}\left(r_g\right) = 2\pi r_g \sin\psi_g \tag{2.24}$$

$$V_l\left(r_g\right) = \frac{1}{3}\pi r_g^3\left(2 - 3\cos\psi_g + \cos^3\psi_g\right) - \frac{1}{3}\pi R^3\left(2 - 3\cos\phi_g + \cos^3\phi_g\right) \tag{2.25}$$

Where ψ_g is half of the central angle of the droplet, ϕ_g is half of the central angle of the contact between the liquid embryo and the particle, r_g is the radius of the liquid embryo, R is the particle radius, and d_g is the center-to-center distance between the liquid embryo and the fine particles. S_{vl} (r_g), S_{ls} (r_g), and L_{lvs} (r_g) are the contact area of liquid embryo and gas phase, the contact area of liquid embryo and fine particles, and the three-phase contact line, respectively.

According to the above two mechanisms, the condensation coefficient of water vapor on the liquid embryo is:

$$C_g\left(\alpha\right) = \alpha_g \frac{p_v}{\sqrt{2\pi m_{vem} k_B T}} S_{vu}\left(r_g\right) \tag{2.26}$$

$$C_g(\beta) = \delta n_1 v \exp\left(-\frac{\Delta G_{\mathrm{dif}}}{k_B T}\right) L_{\mathrm{ivs}}(r_g) \tag{2.27}$$

$$n_1 = \frac{p_v}{v\sqrt{2\pi m_{wm} k_B T}} \exp\left(-\frac{\Delta G_{\mathrm{des}}}{k_B T}\right) \tag{2.28}$$

Where pv is the steam pressure, Pa; m_{wm} is the mass of a single water molecule, kg; T is the steam temperature, K; δ is the average jump distance of a single water molecule, m; v is the surface vibrational frequency of the water molecule, s^{-1}; ΔG_{diff} is the surface diffusion energy of a single water molecule, J/mol; ΔG is the adsorption energy of a single water molecule, J/mol. α is the adhesion coefficient of water molecules on the surface of the liquid embryo. When the water vapor condenses, its value is in the range of 0.04–1 [33]. In order to simplify the calculation, we assume that all the water molecules that impinge on the liquid embryo adhere to the liquid crystal. embryo, so $\alpha_g = 1$. When the gas environment is in a saturated state, the condensation and evaporation of water vapor on the particle surface are in dynamic equilibrium, so:

$$C_g^{\mathrm{sat}}(\alpha) \cdot n_g^{\mathrm{sat}} = E_{g+1}^{\mathrm{sat}}(\alpha) \cdot n_{g+1}^{\mathrm{sat}} \tag{2.29}$$

$$C_g^{\mathrm{sat}}(\beta) \cdot n_g^{\mathrm{sat}} = E_{g+1}^{\mathrm{sat}}(\beta) \cdot n_{g+1}^{\mathrm{sat}} \tag{2.30}$$

In the formulas, the superscript "sat" indicates that the gas environment is in a saturated state, n_g is the number concentration when the liquid embryo is in equilibrium, and the expression is:

$$n_g = n_1 \exp\left[-\frac{W(r_g)}{k_B T}\right] \exp\left[-\frac{W(r_1)}{k_B T}\right] \tag{2.31}$$

In the formula, $W(r_g)$ is the energy required to form liquid embryos [34].

Further derivation of equation (2.31) can be obtained [35]:

$$\frac{n_g^{\mathrm{sat}}}{n_{g+1}^{\mathrm{sat}}} = \exp\left(K_{e_g}\right) \tag{2.32}$$

Where K_{e_g} is the Kelvin number of the liquid embryo, expressed as $K_{e_g} = 2\sigma V_{wm}/(k_B T r_g)$.

Since the evaporation coefficient of water vapor on the particle surface is independent of the water vapor state, therefore:

$$E_g = C_g^{\mathrm{sat}} \cdot \frac{n_g^{\mathrm{sat}}}{n_{g+1}^{\mathrm{sat}}} = \exp\left(K_{e_g}\right) \cdot C_g^{\mathrm{sat}} \tag{2.33}$$

$$E_g = C_g^{eq} = C_g^{eq}(\alpha) + C_g^{eq}(\beta) = \exp\left(\frac{2\sigma V_{wm}}{k_B T r_g}\right) \cdot C_g^{sat} \tag{2.34}$$

To sum up, the growth rate equation of water vapor on the liquid embryo on the particle surface can be obtained:

$$\left.\frac{dr}{dt}\right|_{\text{het}} = \left.\frac{dr}{dg}\right|_{\text{het}} \cdot \left.\frac{dg}{dt}\right|_{\text{het}}$$

$$\left.\frac{dr}{dt}\right|_{\text{het}} \approx \frac{V_{wm}}{\pi r_g^2\left(2-3\cos\psi_g+\cos^3\psi_g\right)} \cdot \left.\left(C_g - E_g\right)\right|_{\text{het}} \tag{2.35}$$

$$n_1 = \frac{p_v}{v\sqrt{2\pi m_{wm} k_B T}}\exp\left(-\frac{\Delta G_{\text{des}}}{k_B T}\right)$$

When the liquid embryo grows to a certain extent, it will completely envelop the fine particles, which can be regarded as a layer of liquid film around the fine particles. At this point, the droplets absorb the surrounding water vapor and continue to grow, and the growth method can be considered as homogeneous condensation. At this stage, we make the following assumptions:

1. Collisions between particles are ignored.
2. The entire droplet is assumed to be at the same temperature, and the temperature gradient within the droplet is ignored.
3. Since the droplet spacing is much larger than the size of the droplet itself, the vapor concentration and temperature in the main gas-phase region between the droplets can be assumed to be uniform, and there is only a concentration gradient and temperature gradient in the gas film layer adjacent to the droplet surface.

By analyzing the heat and mass balance in water vapor condensation on a single particle surface, it can be seen that the total heat flux of the droplet is equal to the sum of the change in the internal energy of the droplet and the energy flowing into the droplet surface:

$$c_{pl} m_d \frac{dT_d}{dt} + h_l I_c = Q \tag{2.36}$$

Assuming that the heat flux from the gas to the droplet is q, the heat is transferred by thermal conduction and condensation diffusion, ignoring the Dufour effect [36]:

$$q = -\lambda \frac{\partial T}{\partial r} + nH_v v \tag{2.37}$$

Where q is the heat flux density, J/m^2/s; λ is the thermal conductivity, W/m/K; n is the number of water molecules, #·m^{-3}; v is the average velocity of condensed vapor molecules, m/s; H is the enthalpy of water vapor, J. Mass transfer density j_v of water vapor molecules flowing into the droplet for:

$$j_v = vnm_{wm} \tag{2.38}$$

Where j_v is the mass transfer density, kg/m²/s.

Put equation (2.38) into equation (2.37); the following expression is obtained: h_v is the specific enthalpy of water vapor, J/ kg:

$$q=-\lambda\frac{\partial T}{\partial r}+j_v\frac{H_v}{m}=-\lambda\frac{\partial T}{\partial r}+j_v h_v \tag{2.39}$$

The growth process of droplets is an unsteady process, but its growth can be approximated as a quasi-steady process [37]. Equation (2.36) can be modified as follows:

$$h_l I_c=Q \tag{2.40}$$

When the heat flow is constant, for spherical particles, it can be obtained:

$$Q=4\pi r^2 q=\text{const} \tag{2.41}$$

Mass flow is expressed as follows:

$$I_c=4\pi r^2 j_v=\text{const} \tag{2.42}$$

The total heat flux into the droplet can be seen as:

$$h_l I_c=-4\pi r^2\lambda\frac{dT}{dr}+I_c h_v \tag{2.43}$$

$$\frac{dT}{dr}=-\frac{I_c(h_l-h_v)}{4\pi r^2\lambda} \tag{2.44}$$

Transform equation (2.44) and integrate both sides to get:

$$\int_{T_\infty}^{T_d}dT=\int_{r_\infty}^{r_d}-\frac{I_c(h_l-h_v)}{4\pi r^2\lambda}dr \tag{2.45}$$

In the formula, T_∞ is the temperature at an infinite radius; r_d is the radius of the droplet; $1/r_\infty$ is 0; (h_l-h_v) is the latent heat of vaporization, denoted by L.

$$T_d=T_\infty-\frac{I_c L}{4\pi r_d\lambda} \tag{2.46}$$

According to the above assumptions, the steady-state mass flow rate is:

$$J_v=-D_v\frac{\partial\rho}{\partial r} \tag{2.47}$$

where D_v is the water vapor diffusion coefficient, ρ is the water vapor density.

Assuming an ideal gas state, equation (2.42) is transformed into equation (2.48), M_v is the molar mass of the vapor, and R^* is the universal gas constant:

$$J_v = -\frac{D_v M_v}{R^* T}\frac{\partial P_v}{\partial r} \tag{2.48}$$

According to equation (2.42), the mass flow rate of the inflow droplet is:

$$I_c = -4\pi r^2 J_v = 4\pi r^2 \frac{D_v M_v}{R^* T}\frac{\partial P_v}{\partial r} \tag{2.49}$$

Transform the above formula into:

$$I_c^{r_d} \int \frac{1}{r^2} dr = -4\pi \frac{D_v M_v}{R^* T} \cdot {}_{P_v}^{P_{v,d}} \int dP_v \tag{2.50}$$

Integrate for r and P_v in equation (2.50), respectively, to get:

$$I_c = \frac{4\pi r_d D_v M_v}{R^* T}\left(P_{v,d} - P_v\right) \tag{2.51}$$

In equation (2.51), $P_{v,d}$ is the partial pressure of vapor at the particle surface, Pa; P_v is the partial pressure of water vapor in the surrounding environment of the fine particles, Pa; when the water vapor condenses on the surface of the particles, $P_v > P_{v,d}$, the mass flow value is negative, so a sign is added to the right side of the formula to meet the calculation requirements:

$$I_c = -\frac{dm_d}{dt} = -4\pi \rho_l r_d^2 \frac{dr_d}{dt} \tag{2.52}$$

Where dm_d/dt is the amount of change in the mass of fine particles over time.

Combine equations (2.51) and (2.52), we can get:

$$r_d dr_d = -\frac{D_v M_v}{\rho_l R^* T}\left(P_{v,d} - P_v\right) dt \tag{2.53}$$

Integrating equation (2.53) gives:

$$\int_{r_{d,0}}^{r_{d,f}} r_d \, dr_d = -\frac{D_v M_v}{\rho_l R^* T}\left(P_{v,d} - P_v\right)_{t0}^{tf} \int dt \tag{2.54}$$

Therefore, the relational expression of droplet radius with time is:

$$r_d = \sqrt{r_{d,0}^2 + \frac{2 D_v M_v}{\rho_l R^+ T_\infty}\left(P_v - P_{v,d}\right) t} \tag{2.55}$$

The vapor partial pressure on the particle surface is related to temperature and supersaturation [1,38], and its expression is:

$$P_{v,d} = P^{\circ}(T_d)exp\left(\frac{2\sigma M_w}{R^* T_d \rho_l r_d}\right) \tag{2.56}$$

Where P^{o} (T_d) is the saturated vapor pressure under the temperature T_d, Pa; σ is the liquid surface tension, mN/m; M_w is the molar mass of the liquid, kg/mol.

The saturated vapor pressure of water vapor at different temperatures is given by the Clapeyron equation [28]:

$$P_{v,d} = P^{\circ}(T_d)exp\left(\frac{2\sigma M_w}{R^* T_d \rho_l r_d}\right)\frac{P^{\circ}(T_d)}{P^{\circ}(T_\infty)} = \exp\left[\frac{LM_w}{R^*}\left(\frac{T_d - T_\infty}{T_d T_\infty}\right)\right] \tag{2.57}$$

Studies have shown that the number concentration of particles in the polydisperse particle groups has great influence on the final particle condensation size [39]. In order to study the growth characteristics of polydisperse particles in a supersaturated water vapor environment, it is crucial to investigate both the droplet growth process and the effect of droplet growth on the surrounding water vapor environment.

The time-discrete model [40] is used to calculate the growth process of polydisperse particle groups. This model assumes that smooth spherical fine particles are uniformly distributed in space, the distribution of vapor concentration gradient and temperature gradient between particles is horizontal, and only high vapor concentration and temperature gradient exist on the particle surface. Therefore, the vapor pressure and temperature in the surrounding continuous phase can be calculated by solving the vapor mass and energy conservation equations. Then, the growth of droplets is calculated by the growth law of a single droplet. The distance between particles is assumed to be much larger than the size of the particles, so particle–particle interactions are ignored. The study by Carstens et al. [41] showed that the effect on the vapor concentration and temperature distribution of the two droplets occurs only when the distance between the two droplets is in the order of the droplet radius. For example, for a particle group with a number concentration of 10^{12} particles/m^3, a particle size of 1 μm, and a uniform distribution, the distance between particles is 100 μm, which is much larger than the particle size, and the possibility of mutual influence between particles is very small. However, in practice, the particles are not completely uniformly distributed, and the interaction between particles cannot be completely excluded, which will bring certain errors to the simulation.

In the simulation, we discretely divide the polydisperse particle group into N groups of particle sizes and consider that the number of particles with the initial particle size r_i in a unit volume is N_i, and the vapor flowing into the particles with the initial particle size r_i, and the mass flow is I_{ci}. This means that the condensation of water vapor on the particle surface will consume a large amount of water vapor, resulting in a decrease in the partial pressure of water vapor in the surrounding environment.

Based on the assumptions of the ideal gas formula, we get:

$$P_v = \frac{R^* T_\infty}{M_v} C \tag{2.58}$$

Where C is the steam mass concentration, which can be obtained by derivation with respect to time t:

$$\frac{dP_v}{dt} = \frac{R^* T_\infty}{M_v} \frac{dC}{dt} \tag{2.59}$$

$$dC = \sum_{i=1}^{N} C_i dm_i \tag{2.60}$$

Where C_i is the number of droplets of radius r_i in each grid; dm_i is the mass variable of a single droplet of radius r_i. Therefore:

$$dm_i = -I_{ci} dt \tag{2.61}$$

Where I_{ci} is the mass flux of the droplet with the radius of r_i.

The mass flux on liquid embryos during the transition phase I_{ci} is equal to the increment of the mass of the liquid embryo, then:

$$dm_i = -I_{ci} dt I_{ci} = \left(C_{gi} - E_{gi}\right) V_L \rho \tag{2.62}$$

Combining equations (2.60) and (2.61), we get:

$$\frac{dP_v}{dt} = -\frac{R^* T_\infty}{M_v} \sum_{i=1}^{N} N_i I_{ci} \tag{2.63}$$

At time t, the water vapor partial pressure P around the droplet is calculated as

$$P_v^{'} = P_{v0} - \frac{R^* T_\infty}{M_v} \sum_{i=1}^{N} N_i I_{ci} t \tag{2.64}$$

Where P_{v0} is the initial partial pressure of water vapor; the latent heat of vaporization will be released when the water vapor condenses on the particle surface, resulting in an increase in the ambient temperature. If it is assumed that the droplet growth process is adiabatic, according to the energy per unit volume, the balance can be obtained [39]:

$$V\left(C_{pg} C_g + C_{pv} C_v\right) \frac{dT}{dt} = LV \sum_{i=1}^{N} I_i C_i \tag{2.65}$$

$$C_g = \left(P - P_v\right) \frac{Mg}{R^* T} \tag{2.66}$$

Where C_{pg} and C_{pv} are the constant pressure specific heat capacities of the gas and water vapor, respectively; M_g is the molar mass of the gas; P is the total pressure of the gas; C_g is the gas mass concentration.

Combining equations (2.66), (2.65), and (2.58), we get:

$$\frac{dT_\infty}{dt} = -\frac{R^* T_\infty L}{(P - P_v)C_{pg}M_g + P_v C_{pv} M_v} \sum_{i=1}^{N} N_i I_{ci} \tag{2.67}$$

According to equation (2.67), the calculation expression of the gas-phase ambient temperature T' around the droplet with time can be obtained:

$$T'_\infty = T_{\infty,0} e^{\alpha t} \tag{2.68}$$

where T_∞,0 is the initial gas-phase temperature:

$$\alpha = \frac{R^* L}{(P - P_v)C_{pg}M_g + P_v C_{pv} M_v} \sum_{i=1}^{N} N_i I_{ci} \tag{2.69}$$

Equations (2.64) and (2.67) give the change of gas-phase pressure caused by the consumption of steam and the change of gas-phase temperature caused by the release of latent heat of vaporization during the growth of the particle group, respectively.

During the transient phase, the growth of the particle population in the growth tube leads to vapor dissipation, resulting in changes in gas-phase pressure and gas-phase temperature in a water vapor environment. In the time-discrete model, the values of pressure and temperature changes are introduced into the condensation rate and evaporation rate of water vapor on the particle surface, and the new condensation rate and evaporation rate can be obtained and then introduced into equation (2.35) to obtain the next growth rate of the liquid embryo on the surface of the node particle, thereby enabling the calculation of the growth of the liquid embryo on the particle surface in the transition stage.

In the growth stage, replace the equation (2.62) with the mass flux equation (2.51) in the growth stage to obtain the variation of pressure and temperature. Then, equations (2.63) and (2.68) are brought into the single droplet to grow the particle. From the calculation equation (2.55) of the diameter, the calculation formula of the grown particle size of the particle group can be obtained:

$$r_d = \sqrt{r_{d,0}^2 + \frac{2D_v M_v}{\rho_l R^* T'_\infty}\left(P'_v - P_{v,d}\right)t} \tag{2.70}$$

2.4 CONDENSATION GROWTH OF FINE PARTICULATE MATTER

2.4.1 Continuous Mechanism

When the saturation and temperature reach a certain threshold, existing particles act as nucleation sites for condensation of particulate matter. The growth rate of the particles depends on the exchange of molecules and heat between the particles and the surrounding medium. When the diameter of the particles is larger than the average free path of the suspended gas, it is referred to as the continuous mechanism, and the transfer of heat and mass is controlled by the diffusion process. For a single-spherical particle, the diffusion rate is governed by the Fick diffusion equation. In spherical coordinates, the general diffusion equation can be expressed as:

$$\frac{\partial n}{\partial t} = D_{vg}\frac{\partial r^2}{r^2\partial r}\left(\frac{\partial n}{\partial r}\right) \tag{2.71}$$

Where n is the molecular concentration of condensed vapor, D_{vg} is the diffusion coefficient of condensed vapor, and t is the time. The condensation process is a dynamic process where the droplet radius constantly changes, and the rate of condensation varies. However, due to the rapid redistribution of vapor concentration around the droplet, it can be approximated as a quasi-static process. Based on such assumption, the steady-state diffusion of condensation can be integrated over the right-hand side of the equation below:

$$\frac{\partial n}{\partial r} = \frac{c_1}{r^2} \tag{2.72}$$

Where c_1 is the constant. According to Fick's law of diffusion, the rate of vapor diffusion onto the droplet is:

$$\varphi = \pi D_p^2 D_{vg}\left(\frac{\partial n}{\partial r}\right)_{r=D_p/2} \tag{2.73}$$

$$c_1 = \frac{\varphi}{4\pi D_{vg}} \tag{2.74}$$

$$\frac{\partial n}{\partial r} = \frac{\varphi}{4\pi D_{vg} r^2} \tag{2.75}$$

On the droplet surface, $n = n_d$. When $r \to \infty$, $n = n_\infty$. So, the following equation is obtained:

$$n_d - n_\infty = \frac{\varphi}{4\pi D_{vg}}\int_{D_p/2}^{\infty}\frac{dr}{r^2} = -\frac{\varphi}{4\pi D_{vg}\, D_p/2} \tag{2.76}$$

Using the assumption of ideal gas, $n = p/kT$.

$$\varphi = 2\pi D_p \frac{D_{vg}}{k}\left(\frac{p_\infty}{T_\infty} - \frac{p_d}{T_d}\right) \tag{2.77}$$

Where k is the Boltzmann constant. p_∞ and p_d are the vapor pressure at $r \to \infty$ and $r = D_p/2$, respectively. T_∞ and T_d are the temperature at $r \to \infty$ and $r = D_p/2$, respectively.

The rate of change in the volume of particulate matter is:

$$\frac{dv}{dt} = v_m\varphi \tag{2.78}$$

Where v is the droplet volume, and v_m is the volume of condensed vapor. The equation describing the growth of particulate matter in a continuous regime can be written as:

$$D_p \frac{dD_p}{dt} = \frac{4D_{vg}v_m}{k}\left(\frac{p_\infty}{T_\infty} - \frac{p_d}{T_d}\right) \tag{2.79}$$

2.4.2 Free Molecular Mechanism

If the particle size is significantly smaller than the average free path of the surrounding vapor, the condensation or evaporation rate of the particle can be determined solely based on the kinetic theory. The molecular flux, which refers to the total number of molecules surrounding the surface of the droplet per unit time, can be mathematically expressed as:

$$\varphi = \pi D_p^2 \left\{\frac{n_\infty c_\infty}{4} - \frac{n_d c_d}{4}\right\} \tag{2.80}$$

Where n_∞ and n_d are the vapor concentration at $r \to \infty$ and $r = D_p/2$, respectively. c_∞ and c_d are the average molecular heat transfer rate at $r \to \infty$ and $r = D_p/2$. The heat transfer rate at the surface of the droplet is expressed as:

$$c_\infty = \left(\frac{8kT_\infty}{\pi m}\right)^{1/2} \tag{2.81}$$

Where m is the mass of the vapor molecule. Using the ideal gas law, we can obtain:

$$\varphi = \frac{\pi D_p^2}{(2\pi mk)^{1/2}}\left[\frac{p_\infty}{T_\infty^{1/2}} - \frac{p_d}{T_d^{1/2}}\right] \tag{2.82}$$

Therefore, an adjustable parameter A is introduced to consider the proportion of surrounding vapor molecules condensed on the droplet,

$$\varphi = \frac{A\pi D_p^2}{(2\pi mk)^{1/2}}\left[\frac{p_\infty}{T_\infty^{1/2}} - \frac{p_d}{T_d^{1/2}}\right] \tag{2.83}$$

The ratio of vapor molecules that actually condense on the droplet surface to the total number of surrounding vapor molecules is defined as the adjustable parameter A. Typically, the condensation factor is determined through experimentation. In the continuum regime, the growth rate of particles is equal to the molecular flux, which is proportional to the diameter of the particles.

2.4.3 Knudsen Mechanism

Two diffusion-controlled boundary cases govern particle growth: the continuum mechanism and the kinetically controlled free molecule mechanism. The kinetically controlled free molecule mechanism also includes a Knudsen transport mechanism.

Additionally, droplet growth via condensation can be computed using the following equation [42]:

$$d_p \frac{d}{dt}(d_p) = \frac{4Dv}{R}\left(\frac{p}{T} - \frac{p_d}{T_d}\right) f(Kn) \tag{2.84}$$

Where D is the diffusion coefficient of the condensing vapor; p and T are the vapor pressure and temperature in the surrounding gas far away from the particle; and v is the molar volume. The Fuchs correction, $f(Kn)$, is important for particles smaller than 0.1 μm:

$$f(Kn) = \frac{1 + Kn}{1 + 1.71Kn + 1.333Kn^2} \tag{2.85}$$

Where $Kn(= 2\lambda / d_p)$ is the Knudsen number; and λ and d_p are the mean free path of the gas medium and particle diameter, respectively. Furthermore, p_d and T_d are the vapor pressure and temperature at the surface, respectively. For the Kelvin effect, p_d can be calculated from the following equation:

$$\frac{p_d}{p_s} = \exp\left(\frac{4v\gamma}{RTd_p}\right) \tag{2.86}$$

Where γ is the surface tension; v is the molar volume of the liquid; R is the gas constant. The temperature on the droplet surface, T_d, includes the effect of the droplet temperature increase due to the latent heat of condensation and can be estimated as follows [43]:

$$T_d = T + \frac{LMD}{k_v R}\left(\frac{p_\infty}{T_\infty} - \frac{p_d}{T_d}\right) \tag{2.87}$$

Where L is the latent heat of the working fluid; M is the molecular weight of the vapor; k_v is the thermal conductivity of the carrier gas; and p_∞ and T_∞ are the vapor pressure and temperature away from the droplet.

The principle of a CPC involves exposing aerosols to a supersaturated vapor and subsequently cooling them, leading to particle growth via heterogenous nucleation. However, the process of growth through condensation necessitates the use of both a saturator and a condenser, contributing to the complexity of the system. The nucleation processes rely heavily on the saturation rate and temperature, requiring accurate quantitative experimental specification. While hygroscopic particles may take up water at a relative humidity below 100%, small particles generally need to be exposed to a region of vapor supersaturation for their growth through condensation. This is due to the influence of surface energy associated with the curvature of the droplet surface, which depends on the particle size. For bulk liquids, the vapor pressure at the saturation or equilibrium condition varies with temperature. Adiabatic expansion, turbulent mixing, and non-isothermal diffusion are the three main techniques used in the past to establish supersaturated vapor conditions in heterogeneous nucleation studies (Table 2.2).

TABLE 2.2
Commercial Condensation Particle Counter

Company	Type	Working Fluid	Detection Limit (nm)
KAN	Mixing	Propylene glycol	10
GPI	Conductive cooling	Butanol	4.5
TSI	Conductive cooling	Butanol	10
TSI	Conductive cooling	Butanol	10
TSI	Conductive cooling	Butanol	4
TSI	Conductive cooling	Butanol	2.5
TSI	Differential diffusion	Water	5
TSI	Differential diffusion	Water	2.5

2.5 TYPE OF CONDENSATION GROWTH

2.5.1 Adiabatic Expansion-Type Condensation Growth

The concept of adiabatic expansion condensation growth can be traced back to Espy's invention of the "cloud" formation instrument (nephelescope) in 1841 [44]. The adiabatic expansion was achieved by releasing the stopcock in a pressurized vessel containing aerosols. In 1875, Coulier demonstrated that condensation occurs more readily in unfiltered air with aerosols during adiabatic expansion, as compared to clean air [16]. This finding is the basis for particle detection in the gas phase using CPCs.

Another significant contribution to adiabatic expansion condensation was made by Wilson in 1897 [45]. Wilson emphasized the importance of controlling the expansion ratio to eliminate ion-induced and homogenous nucleation. He discovered that cloud formation does not occur until the volume expansion ratio reaches 1.252 after removing particles. Furthermore, Wilson observed that a dense blue or blue-green fog appears for volume expansion ratios between 1.41 and 1.42, which becomes white for ratios exceeding 1.44.

The principle of adiabatic expansion-type condensation growth can be divided into two steps: firstly, increasing the humidity of the aerosol system in a closed vessel to achieve water vapor saturation; secondly, rapidly cooling the system by volume expansion or pressure release. This prompts the aerosol system to shift toward a supersaturation state at a lower ambient temperature, leading to the condensation of water vapor onto fine particles through heterogeneous nucleation.

Aitken designed an adiabatic expansion-type CPC prototype in 1888 [17], which used piston motion to expand the air and a valve to release the pressure in the high-pressure system. Figure 2.5 shows a schematic diagram of adiabatic expansion-type CPCs. The expansion ratio (ε_r) is determined by the volume or pressure ratio and can be defined as:

$$\varepsilon_r = \frac{p_i}{p_f} = \frac{v_f}{v_i} \tag{2.88}$$

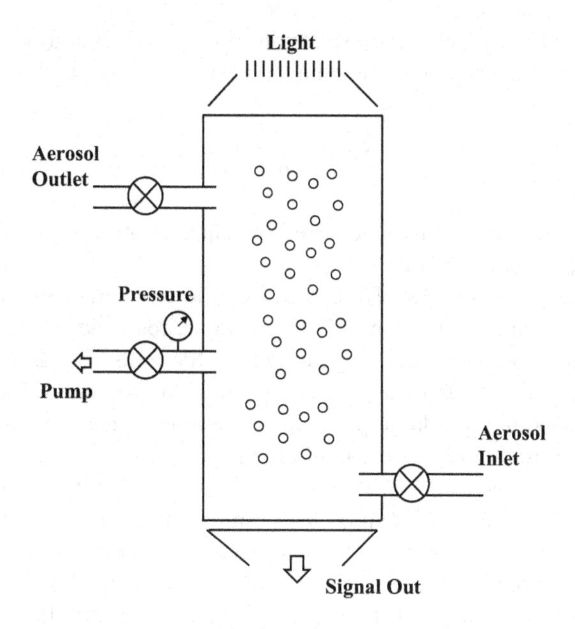

FIGURE 2.5 Schematic diagram of adiabatic expansion type.

Where p_i, p_f, v_i, and v_f denote the initial pressure, final pressure after expansion, initial volume, and final volume after expansion, respectively. The ε_r in typical CPC is between 1.1 and 1.5. If we assume that the ideal adiabatic expansion occurs in the water vapor and air mixture, then the following equation can be derived:

$$\frac{p_i}{p_f} = \left(\frac{v_f}{v_i}\right)^{\gamma} \tag{2.89}$$

Where γ is the isentropic exponent, for example, the γ of air is about 1.4. After expansion, the relation between initial and final temperature can be calculated by ε_r and γ:

$$\frac{T_f}{T_i} = \left(\frac{1}{\varepsilon_r}\right)^{\gamma-1} \tag{2.90}$$

Saturation rate S_R is defined as the ratio of the partial pressure $p(T_f)$ of water to the saturated water pressure $p_s(T_f)$, as shown in the equation below:

$$S_R = \frac{p(T_f)}{p_s(T_f)} \tag{2.91}$$

The temperature after expansion T_f determines the corresponding equilibrium vapor pressure $p(T_f)$ and enables the calculation of S_R by considering the changing partial vapor pressure in the course of the expansion. If the initial condition is under water vapor saturation condition, then after adiabatic expansion, S_R can denoted as:

$$S_R = \varepsilon_r \frac{p_s(T_i)}{p_s(T_f)} \tag{2.92}$$

In this way, a desirable S_R can be achieved by adjusting ε_r to make a condensation growth on fine particulate matter.

During the cooling process of the aerosols, the temperature of the expansion chamber walls remains under controlled conditions, resulting in an instantaneous increase in the total pressure due to reheating from the walls. This issue can be minimized by using pressure-defined expansion instead of volume-defined expansion, achieved by connecting to a large pressure vessel at low pressure via an expansion valve. This configuration allows the pressure increase from the walls to be buffered by a large volume, ensuring that it does not affect the conditions in the measuring volume for extended periods until measurements are conducted. As a result, the obtained saturation ratios exhibit remarkable accuracy, making the expansion chamber an ideal instrument for heterogeneous nucleation studies [46]. This method has been used to investigate heterogeneous nucleation on neutral particles with a diameter as small as 1.4 nm and ions in the sub-1 nm range [47]. However, the expansion process requires a specific sequence of steps for flushing and equilibration of the aerosol with the chamber thermal settings, limiting the time response to changing aerosols. Measurements can be taken at intervals of approximately 1 minute. This limitation is negligible under defined laboratory conditions of constant particle properties. One significant drawback of the expansion-type CPC is its cyclic flow, which makes it incompatible with the steady-state flow requirement when used to measure concentration from an electrical mobility analyzer.

2.5.2 Turbulent Mixing-Type Condensation Growth

Supersaturated vapor conditions can be generated through the turbulent mixing of warm, saturated gas with cold aerosols, eliminating the need for auxiliary cooling devices. This process leads to heterogeneous nucleation, which is often observed in wintertime when exhaled air produces small clouds in front of a person. The highly turbulent flow creates a quasi-uniform mixture with negligible particle loss and rapid condensation growth. This method has been optimized to reduce the total response time to 0.45 seconds, outperforming the TSI UCPC Model 3025's response time of 2.7 seconds. It achieves a minimum detection diameter of 5 nm, enabling rapid-scanning differential mobility analyzer measurements.

The turbulent mixing-type condensation growth technique for nanoparticle detection was first introduced by Okuyama et al. [20]. Probably the first quantitative study of heterogeneous nucleation on sub-1 nm ions was carried out by this method [48]. This method was modified by using working fluids with low saturation vapor pressures to obtain saturation ratios around 1,000. However, the low vapor concentration limits the size of particles activated by heterogeneous nucleation, making them too small for optical detection. To solve this problem, a booster stage was introduced to detect particles in a second CPC stage. This method is now commercially available and capable of detecting particles as small as 1 nm [49]. The technique allows for the detection of extremely small condensation nuclei due to the high saturation ratios applied and continuous flow operation, providing time resolution on the order of a few seconds. The saturator flow rate can be varied to determine the size distribution in the sub-3 nm range. However, the turbulent mixing technique lacks precision in determining saturation ratios and does not provide a defined region of uniform saturation. Therefore, a quantitative test of nucleation theory is difficult to perform using this method.

Usually, the turbulence is generated by an impinging jet flow, and the mixing time scale (τ_t) between two warm and cold streams in a circular tube can be described by the turbulent dissipation rate:

$$\tau_t = \left[2N_j^2 \left(\frac{r_j}{r} \right)^4 \left(\frac{r}{L} \right)^2 \left(\frac{1}{\varnothing N_j} \right)^2 \right]^{\frac{1}{3}} \tag{2.93}$$

Where L and r is the length and radius of the circular tube, r_j is the injector nozzle diameter and N_j is the volume flow rate in each injected pulse. When the injector is small enough, the residence time τ_r is longer than the mixing time τ_t, allowing the two streams to mix thoroughly. However, if the injector is too large, the turbulence will decay in the downstream of the circular tube, and mixing will continue. Figure 2.6 shows a schematic diagram of the turbulent mixing-type condensation growth, which requires cold aerosol and warm saturated gas to enter the mixing chamber simultaneously. After mixing, the mixture enters the condensation chamber, where it quickly condenses on the surface of particulate matter in a cold condenser, promoting further condensation growth. The flow in the mixing chamber is highly turbulent, making it difficult to accurately describe temperature and pressure distribution. Assuming ultra-fast mixing without wall heat transfer loss, the following equations are derived based on heat and mass conservation laws:

$$T_{\text{mix}} = \frac{\left(C_{\text{sat}} T_{\text{sat}} Q_{\text{sat}} + C_a T_a Q_a \right)}{C_{\text{mix}}} \tag{2.94}$$

$$H_{\text{mix}} = \frac{Q_{\text{sat}} H_{\text{sat}}}{Q_{\text{mix}}} \tag{2.95}$$

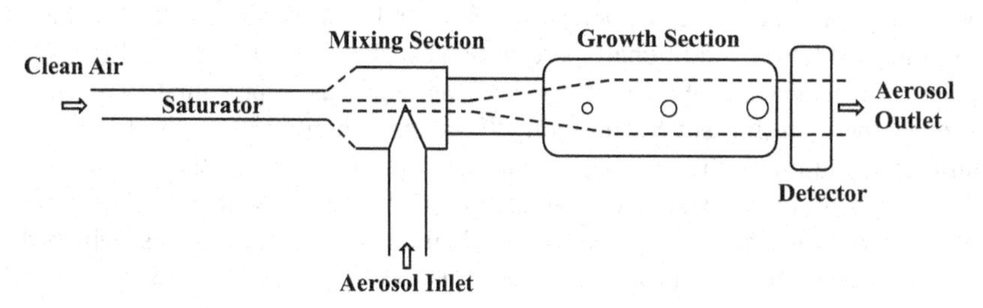

FIGURE 2.6 Schematic diagram of turbulent mixing type.

Where C, Q, and H are the heat capacity, flow rate, and the amount of saturation water vapor, and the subscript "sat", "a", and "mix" denote saturation stream, aerosol stream, and mixed stream:

$$H = \frac{M_{\text{fluid}}}{M_{\text{air}}} \frac{p}{p_{\text{atm}} - p} \tag{2.96}$$

2.5.3 Non-isothermal Diffusion-Type Condensation Growth

The most commonly used approach for producing supersaturated vapor in CPCs involves non-isothermal diffusion. In laminar flow devices, the aerosol is introduced to a cylindrical saturator at a fixed temperature, allowing vapor to diffuse into the aerosol. The saturated aerosol flow is then quickly cooled or heated in a condenser tube, which is maintained at a temperature lower or higher than the saturator, depending on the chosen working fluid. The direction of the temperature gradient is determined by the diffusion properties of the vapor relative to the carrier gas. For many applications, a continuous flow CPC that employs water as the working fluid is desirable to avoid issues such as exposure to alcohol fumes in indoor environments and accidental spills. However, diffusional cooling CPCs that utilize water as the condensing vapor encounter limitations due to the high mass diffusivity of water vapor. Specifically, the mass diffusivity of water vapor is $2.4 \times 10^{-5}\ \text{m}^2/\text{s}$ at 20°C, which is greater than the thermal diffusivity of air at the same temperature. As a result, in a thermal diffusion cooling-type CPC, water vapor diffuses to the walls of the condenser faster than the flow cools. This depletes the water vapor and leads to a lower degree of supersaturation compared to simply mixing hot and cool saturated airstreams. However, for working fluids with large molecules such as n-butanol, the heat diffusivity of the carrier gas (typically air) is much faster than the mass diffusivity of butanol vapor. Therefore, the condenser is maintained at a lower temperature than the saturator, aiming to obtain supersaturation by reducing $p_s(T_f)$.

When water is used as the working fluid in CPCs, a reversed temperature gradient (reversed T-gradient) is essential due to the faster diffusion of water molecules compared to air molecules, owing to their lighter weight. Consequently, the warmer condenser is equipped with wetted walls, which allows additional water molecules to diffuse into the saturated aerosol flow. Since the heat diffusivity is slower, the supersaturation increases due to an increase in the vapor pressure p_v [50]. The non-isothermal diffusion method, which involves a simple straight tube and continuous flow mode providing time resolution of less than a second, has become the leading method in commercial CPCs. Additionally, precautions such as sheathed aerosol flows have enabled the nucleation of neutral particles as small as approximately 2 nm [51]. However, the resulting supersaturation (S) values exhibit a parabolic profile, with saturated conditions at the walls ($S = 1$) and maximum S values in the central streamline. Depending on the size and composition of the particles, they will experience substantially different regions of S along their respective trajectories, which may or may not activate their growth. The use of non-isothermal diffusion devices in quantitative heterogeneous nucleation studies has been limited in the past, but recent

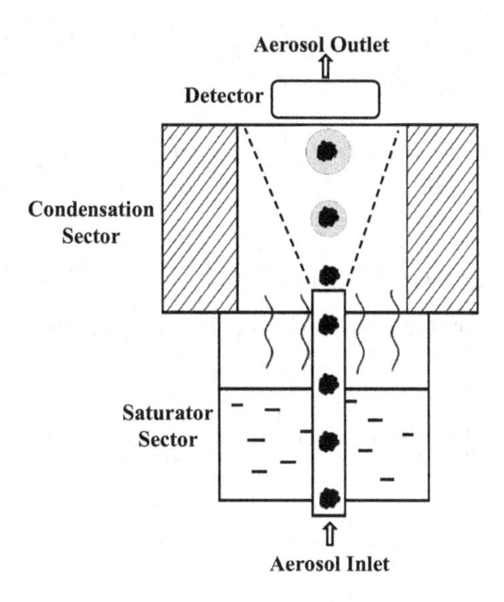

FIGURE 2.7 Schematic diagram of non-isothermal diffusion type.

progress by the Fernandez de la Mora group suggests that sheathing the saturated aerosol flow may provide sufficiently uniform S values and open up new possibilities for such experiments [52].

Figure 2.7 depicts a schematic of a non-isothermal diffusion-type system. In order to accurately count particles with an optical particle counter, a segment of the aerosol flow was first filtered to reduce concentrations. Subsequently, the aerosols then passed through an annular gap where the outer walls were cooled with an ice bath while the temperature-controlled center rod was kept moist and heated. This temperature gradient caused supersaturation of water vapor, leading to the condensational growth of particles before detection by the optical particle counter.

To determine the activation efficiency of particles of varying sizes, it is crucial to calculate the supersaturation at all points in the growth cavity using the saturation formula. This requires knowledge of the temperature T and steam partial pressure p_v at each location in the chamber. As such, it is necessary to first calculate the distribution of steam partial pressure and temperature in the growth chamber.

If the particulate matter gas stream satisfies the following assumptions:

1. The fluid is an incompressible Newtonian fluid;
2. Ignore the heat generated by axial diffusion and friction or pressure changes;
3. The fluid moves in a laminar flow state in the cavity and the flow rate boundary layer is fully developed before entering the cavity, and the airflow velocity profile is parabolic:

$$v_z(r) = v_0\left(1-\frac{r^2}{R^2}\right) \tag{2.97}$$

then the differential equation for heat transfer in the one-dimensional layer of the circular tube can be simplified:

$$\rho C_p\left(\frac{\partial T}{\partial t}+v_r\frac{\partial T}{\partial r}+\frac{v_\theta}{r}\frac{\partial T}{\partial \theta}+v_z\frac{\partial T}{\partial z}\right)=k\left[\frac{1}{r}\frac{\partial}{\partial r}\left(r\frac{\partial T}{\partial r}\right)+\frac{1}{r^2}\frac{\partial^2 T}{\partial \theta^2}+\frac{\partial^2 T}{\partial z^2}\right]+\mu\Phi_v \quad (2.98)$$

As a steady-state fluid flows through a pipe $\left(\frac{\partial T}{\partial r}=0,\frac{\partial p_v}{\partial r}=0\right)$, the distribution of its temperature and pressure fields can be described by the energy equation for Newtonian fluids:

$$v_0\left(1-\frac{r^2}{R^2}\right)\frac{\partial T}{\partial z}=\alpha\left[\frac{1}{r}\frac{\partial}{\partial r}\left(r\frac{\partial T}{\partial r}\right)+\frac{\partial^2 T}{\partial z^2}\right] \quad (2.99)$$

$$v_0\left(1-\frac{r^2}{R^2}\right)\frac{\partial p_v}{\partial z}=D_v\left[\frac{1}{r}\frac{\partial}{\partial r}\left(r\frac{\partial p_v}{\partial r}\right)+\frac{\partial^2 p_v}{\partial z^2}\right] \quad (2.100)$$

Where $\alpha=\frac{k}{\rho C_p}$ is the thermal diffusion coefficient, D_v is the vapor mass diffusion coefficient, and R is the radius of the tube. The boundary conditions of temperature and pressure are:

1. the tube inlet: $T(r,0)=T_0, p_v(r,0)=P_0, f(r,0)=f_0$;
2. the tube outlet: $T(R,z)=T_w, p_v(R,z)=P_w$, *i.e.* $f(R,z)=f_w$;
3. the tube central line: $\frac{\partial T}{\partial r}(0,z)=0$.

Once the inlet, saturation chamber, and condensing chamber wall surface boundary conditions are defined, equation (2.98) can be solved using the separation variable method to obtain the distribution of the temperature and vapor pressure fields. The ratio of vapor pressure at a given point in the tube to the saturated vapor pressure of the working fluid defines the degree of supersaturation at that point. The Kelvin particle size, which represents the critical activation particle size of particulate matter, can also be calculated at any point in the condensation chamber. Finally, using the forced convective condensation growth model, the activation efficiency of particulate matter can be determined.

2.6 CONCLUSIONS

This chapter conducts an analysis of the heat and mass transfer processes of fine particles in a supersaturated environment. It investigates the dynamic processes of single particles during the nucleation stage, transition stage, and growth stage, providing formulas for the growth rate of liquid droplet nuclei during the transition stage and the particle growth during the growth stage. By analyzing the vapor consumption caused by particle growth and the release of latent heat of vaporization during particle evaporation, the variations in vapor phase pressure and vapor phase temperature in the water vapor environment are determined. Consequently, the growth formulas for particle clusters during the transition and growth stages are obtained. In addition, this chapter introduces the continuous mechanism, free molecular mechanism, and Knudsen mechanism for the growth of heterogeneous condensation of particulate matter and also gives three ways of creating a supersaturated vapor atmosphere, namely, expansion, turbulent mixing and non-isothermal diffusion, and analyses the respective advantages and disadvantages of the three ways and their thermodynamic process. This chapter provides a detailed discussion of the particulate matter condensation growth process from a theoretical modeling perspective.

REFERENCES

[1] Abraham FF, Zettlemoyer AC. Homogeneous nucleation theory. *Physics Today.* 1974, 27(12): 52–3.
[2] Wagner W, Cooper JR, Dittmann A. The IAPWS industrial formulation 1997 for the thermodynamic properties of water and steam. *Journal of Engineering for Gas Turbines and Power.* 2000, 122: 151.
[3] Fan Y, Qin F, Luo X, Lin L, Gui H, Liu J. Heterogeneous condensation on insoluble spherical particles: Modeling and parametric study. *Chemical Engineering Science.* 2013, 102: 387–96.
[4] Liu R. *The Nucleation and Growth Characteristics of Fine Particles in the Process of Vapor Heterogeneous Condensation.* East China University of Science and Technology, Shanghai, 2019.
[5] Xu J-C. *Characteristics of Coal-Fired PM2.5 Nucleation and Growth under Supersaturated Water Vapor.* Southeast University, Nanjin, 2018.
[6] Yin P, Li T, Cao X, Teng L, Li Q, Bian J. Condensation properties of water vapor under different back pressures in nozzle. *Case Studies in Thermal Engineering.* 2022, 31: 101783.
[7] Guo D, Cao X, Zhang P, Ding G, Liu Y, Cao H, Bian J. Heterogeneous condensation mechanism of methane-hexane binary mixture. *Energy.* 2022, 256: 124627.
[8] Fletcher NH. Size effect in heterogeneous nucleation. *The Journal of Chemical Physics.* 1958, 29(3): 572–6.
[9] Talanquer V, Oxtoby DW. Nucleation on a solid substrate: A density functional approach. *The Journal of Chemical Physics.* 1996, 104(4): 1483–92.
[10] Wang ZJ, Wang SY, Wang DQ, Yang YR, Wang XD, Lee DJ. Water vapor condensation on binary mixed substrates: A molecular dynamics study. *International Journal of Heat and Mass Transfer.* 2022, 184: 122281.
[11] Fil B E, Kini G, Garimella S. A review of dropwise condensation: Theory, modeling, experiments, and applications. *International Journal of Heat and Mass Transfer.* 2020, 160: 120172.
[12] Derjaguin BV, Prokhorov AV. Improved theory of homogeneous condensation and its comparison with experimental data. *Journal of Colloid and Interface Science.* 1974, 46(2): 283–5.
[13] Kashchiev D. *Nucleation: Basic Theory with Applications.* Butterworth Heinemann, 2001.
[14] Gorbunov B. Free energy of embryo formation for heterogeneous multicomponent nucleation. *The Journal of Chemical Physics.* 1999, 110(20): 10035–45.
[15] Davis, EJ. Transport phenomena with single aerosol particles. *Aerosol Science and Technology.* 1982, 2(2): 121–44.

[16] Wilson, CTR. XI. Condensation of water vapour in the presence of dust-free air and other gases. *Philosophical Transactions of the Royal Society of London.* Series A, Containing Papers of a Mathematical or Physical Character, 1897 (189): 265–307.
[17] Aitken J. I.—On the number of dust particles in the atmosphere. Earth and *Environmental Science Transactions of the Royal Society of Edinburgh.* 1889, 35(1): 1–9.
[18] Hope T. *Transactions of the Royal Society of Edinburgh,* Edinburgh, Scotland, 1905.
[19] Wilson CT. On an expansion apparatus for making visible the tracks of ionising particles in gases and some results obtained by its use. *Proceedings of the Royal Society of London. Series A, Containing Papers of a Mathematical and Physical Character.* 1912, 87(595): 277–92.
[20] Okuyama K, Kousaka Y, Motouchi T. Condensational growth of ultrafine aerosol particles in a new particle size magnifier. *Aerosol Science and Technology.* 1984, 3(4): 353–66.
[21] Moran MJ, Shapiro HN, Boettner DD, et al. *Fundamentals of engineering thermodynamics.* John Wiley & Sons, New York, 2010.
[22] Midya US, Bandyopadhyay S. Operation of Kelvin effect in the activities of an antifreeze protein: A molecular dynamics simulation study. *The Journal of Physical Chemistry B.* 2018, 122(12): 3079–87.
[23] Thomson W. LX. On the equilibrium of vapour at a curved surface of liquid. *The London, Edinburgh, and Dublin Philosophical Magazine and Journal of Science.* 1871, 42(282): 448–52.
[24] Gibbs JW. On the equilibrium of heterogeneous substances. *American Journal of Science.* 1878, 3(96): 441–58.
[25] Ahn KH, Liu BY. Particle activation and droplet growth processes in condensation nucleus counter—II. Experimental study. *Journal of Aerosol Science.* 1990, 21(2): 263–75.
[26] Stolzenburg MR, McMurry PH. An ultrafine aerosol condensation nucleus counter. *Aerosol Science and Technology.* 1991, 14(1): 48–65.
[27] Ahn KH, Liu BY. Particle activation and droplet growth processes in condensation nucleus counter—I. Theoretical background. *Journal of Aerosol Science.* 1990, 21(2): 249–61.
[28] Steinfeld JI. Atmospheric chemistry and physics: From air pollution to climate change. *Environment: Science and Policy for Sustainable Development.* 1998, 40(7): 26.
[29] Emeritus ZB. Food Process Engineering and Technology (Second Edition), Israel Institute of Technology, Haifa, Israel, 2013, pp. 353–71.
[30] Gurav A, Kodas T, Pluym T, et al. Aerosol processing of materials, Aerosol Science and Technology, 1993, 19(4): 411–452.
[31] Bolsaitis PP, McCarthy JF, Mohiuddin G, Elliott JF. Formation of metal oxide aerosols for conditions of high supersaturation. *Aerosol Science and Technology.* 1987, 6(3): 225–46.
[32] Fan Y, Qin F, Luo X, Zhang J, Wang J, Gui H, Liu J. A modified expression for the steady-state heterogeneous nucleation rate. *Journal of Aerosol Science.* 2015, 87: 17–27.
[33] Laaksonen A, Vesala T, Kulmala M, Winkler PM, Wagner PE. Commentary on cloud modelling and the mass accommodation coefficient of water. *Atmospheric Chemistry and Physics.* 2005, 5(2): 461–4.
[34] Girshick SL, Chiu CP. Kinetic nucleation theory: A new expression for the rate of homogeneous nucleation from an ideal supersaturated vapor. *The Journal of Chemical Physics.* 1990, 93(2): 1273–7.
[35] Luo X, Fan Y, Qin F, Gui H, Liu J. A kinetic model for heterogeneous condensation of vapor on an insoluble spherical particle. *The Journal of Chemical Physics.* 2014, 140(2): 1873.
[36] Heidenreich S. Condensational droplet growth in the continuum regime—A critical review for the system air-water. *Journal of Aerosol Science.* 1994, 25(1): 49–59.

[37] Fuchs NA, Sabersky RH, Pratt JN. Evaporation and droplet growth in gaseous media. *Journal of Applied Mechanics*. 1960, 27(4): 55.
[38] Mcdonald JE. Homogeneous nucleation of vapor condensation. I. Thermodynamic aspects. *American Journal of Physics*. 1962, 30(12): 225–37.
[39] Heidenreich S, Ebert F. Condensational droplet growth as a preconditioning technique for the separation of submicron particles from gases. *Chemical Engineering and Processing: Process Intensification*. 1995, 34(3): 235–44.
[40] Girshick SL, Chiu CP, McMurry PH. Time-dependent aerosol models and homogeneous nucleation rates. *Aerosol Science and Technology*. 1990, 13(4): 465–77.
[41] Carstens JC, Williams A, Zung JT. Theory of droplet growth in clouds: II. Diffusional interaction between two growing droplets. *Journal of Atmospheric Sciences*. 1970, 27(5): 798–803.
[42] Fuchs NA, Sutugin AG. Coagulation rate of highly dispersed aerosols. *Journal of Colloid Science*. 1965, 20(6): 492–500.
[43] Zhang ZQ, Liu BY. Dependence of the performance of TSI 3020 condensation nucleus counter on pressure, flow rate, and temperature. *Aerosol Science and Technology*. 1990, 13(4): 493–504.
[44] Espy JP. *The Philosophy of Storms*. CC Little and J. Brown, Boston, MA, 1841.
[45] Wilson CT. Condensation of water vapour in the presence of dust-free air and other gases. *Philosophical Transactions of the Royal Society of London. Series A, Containing Papers of a Mathematical or Physical Character*. 1897, 189: 265–307.
[46] de la Mora JF. Heterogeneous nucleation with finite activation energy and perfect wetting: Capillary theory versus experiments with nanometer particles, and extrapolations on the smallest detectable nucleus. *Aerosol Science and Technology*. 2011, 45(4): 543–54.
[47] Tauber C, Chen X, Wagner PE, Winkler PM, Hogan Jr CJ, Maißer A. Heterogeneous nucleation onto monoatomic ions: Support for the Kelvin-Thomson theory. *ChemPhysChem*. 2018, 19(22): 3144–9.
[48] Seto T, Okuyama K, De Juan L, Fernández de la Mora J. Condensation of supersaturated vapors on monovalent and divalent ions of varying size. *The Journal of Chemical Physics*. 1997, 107(5): 1576–85.
[49] Vanhanen J, Mikkilä J, Lehtipalo K, Sipilä M, Manninen HE, Siivola E, Petäjä T, Kulmala M. Particle size magnifier for nano-CN detection. *Aerosol Science and Technology*. 2011, 45(4): 533–42.
[50] Hering SV, Stolzenburg MR, Quant FR, Oberreit DR, Keady PB. A laminar-flow, water-based condensation particle counter (WCPC). *Aerosol Science and Technology*. 2005, 39(7): 659–72.
[51] Wimmer D, Lehtipalo K, Franchin A, Kangasluoma J, Kreissl F, Kürten A, Kupc A, Metzger A, Mikkilä J, Petäjä T, Riccobono F. Performance of diethylene glycol-based particle counters in the sub-3 nm size range. *Atmospheric Measurement Techniques*. 2013, 6(7): 1793–804.
[52] de la Mora JF. Viability of basic heterogeneous nucleation studies with thermally diffusive condensation particle counters. *Journal of Colloid and Interface Science*. 2020, 578: 814–24.

3 Optical Particle Measurement Theory and Method

Xiaoyan Ma, Liuyong Chang, and Longfei Chen

3.1 LIGHT SCATTERING THEORY

When a beam of light is irradiated on an object, the charged particles, electrons and protons in the material vibrate under the action of the incident electromagnetic wave, and the charges accelerated by the electromagnetic wave radiate electromagnetic waves in all directions. This process is called "secondary radiation". A simple formula can be used to describe the scattering process of light:

$$\text{Scattering} = \text{excitation} + \text{re–excitation} \tag{3.1}$$

The re-radiation of the illuminated material is light scattering. In addition to radiating electromagnetic energy outward through re-radiation, the excited charge will also convert part of the incident electromagnetic energy into other forms of energy, such as thermal energy and chemical energy. This is an absorption process related to scattering.

The average distance between gas molecules under standard temperature and pressure is tens of angstroms, and the distance between solid and liquid molecules is the order of several angstroms or tens of angstroms. Their average distance is several orders of magnitude smaller than the wavelength of visible light, that is, several thousand angstroms. From the light wave scale, solids, liquids, and most gases belong to continuous non-uniform dense media. The scattering of incident electromagnetic waves by such a system is directly related to its microheterogeneity. In the above system, in addition to the effect of the incident field, the molecule is also affected by the secondary field generated by other surrounding molecules. The secondary field is related to the incident field acting on the molecule. Therefore, to solve the light scattering, it is necessary to solve the multi-body electromagnetic scattering related to each other among molecules. Generally, the average field is used to replace the complex field formed by the multi-body to solve the multi-body scattering problem.

The classical theory of light scattering involves the polarization of molecules. The polarization of molecules has many polarization types, such as orientation polarization, displacement polarization, space charge polarization, and thermal relaxation polarization. It can be roughly divided into two categories: one is the

DOI: 10.1201/9781003423195-3

polarization induced by the electric field, and the second is the polarization caused by the reorientation of molecules with electric dipole moments under the action of the electric field. In the first category, the polarization of electrons and ions is caused by the action of the electric field. The polarization of electrons is the result of the distortion of the electron cloud caused by the action of the electric field, thus changing the electron motion orbit. Ion polarization is the relative position variation of positive and negative ions under the action of electric field, which causes the deviation of positive and negative charge centers and generates induced electric dipole moments. The orientation of molecules with inherent electric dipole moments under the action of electric field has a Smooth process. Therefore, it shows the basic characteristics of frequency dependence and dielectric loss. It is worth pointing out that the orientation process of polar molecules is sometimes completed by the action of defects. The absence of jumps around the impurity corresponds to the rotation of the dipole. Some references and related books on the theory of light scattering can refer to [1–11].

3.1.1 Polarization Theory

3.1.1.1 Polarization of Electrons

The electric field acts on the electron to change their orbital motion, so the electron cloud is distorted, leading to the polarization of the electron. The electric dipole moment can be expressed as:

$$p = \alpha_e E_L \tag{3.2}$$

Among which, α_e is the electronic polarizability, E_L is the local electric field of the electron, and m is the mass. The induced electric dipole moment generated by a pair of dipoles with +e and −e charges under the action of the alternating electric field is:

$$p = -\mathrm{eE} = \frac{e^2}{m} \cdot \frac{1}{\omega_0^2 - \omega^2} \cdot E_0 e^{-i\omega t} \tag{3.3}$$

In which, $\omega_0 = \sqrt{\frac{K}{m}}$, and it is the natural vibration angular frequency, K is the stiffness coefficient, so the static polarization is:

$$\alpha_0 = \frac{e^2}{m\omega_0^2} = \frac{e^2}{K} \tag{3.4}$$

If there are multiple pairs of charges, the dipole moment should be summed for each pair:

$$p = \sum_i e_i r_i \tag{3.5}$$

Where r_i is the position vector of the ith charge e_i.

3.1.1.2 Polarization of Dielectric

The polarization of a solid (*P*) is equal to the electric dipole moment *p* per unit volume *V*:

$$P = \frac{p}{V} \tag{3.6}$$

If the solid refers to a system with *N* atoms, then the expression becomes:

$$P = \sum_i N_i \alpha_i E_{Li} \tag{3.7}$$

Where N_i is the polarization rate α_i per unit volume, and E_{Li} is the local field of the ith atom. It is known from solid physics that the local field E_L in cubic crystal medium has the following relationship:

$$E_L = E + \frac{4\pi}{3} P \tag{3.8}$$

Where *E* is the external field, that is, the incident light field, which is the spatial average field inside the medium. $4\pi P/3$ is the additional field generated by the polarization of other atoms in the medium on the atoms, also known as the Lorentz field. According to the following definitions,

$$\varepsilon = \frac{E + 4\pi P}{E} = 1 + \frac{\chi}{\varepsilon_0} \tag{3.9}$$

$$\chi = (\varepsilon - 1)\varepsilon_0 \tag{3.10}$$

$$\frac{P}{E} = \frac{E_L \sum_i N_i \alpha_i}{E_L - \frac{4\pi}{3} P} = \frac{\sum_i N_i \alpha_i}{1 - \frac{4\pi}{3} \sum_i N_i \alpha_i} = \frac{\varepsilon - 1}{4\pi} = \chi \tag{3.11}$$

the Clausis–Mossotti relation is obtained by calculation:

$$\frac{4\pi}{3} \sum_i N_i \alpha_i = \frac{\varepsilon - 1}{\varepsilon + 2} \tag{3.12}$$

Equation (3.12) reflects the relationship between the dielectric constant of the solid and the electron polarization. When the incident frequency is low, both the electron polarization and the ion polarization contribute to the polarization of the solid. When the incident frequency is high, for example, if the laser is incident on the medium, due to small mass of the electron, the movement of the electron follows the change of the optical frequency; while the ion mass is large, so the inertia, and the motion cannot keep up with the change of optical frequency. Therefore, the contribution of ion polarization to the polarization of solid is small. In this case, the polarization of solid mainly comes from the electronic polarization. In fact, the electronic polarization is still quite large in the ultraviolet range, and the frequency range of the large contribution of ion polarization is generally in the long infrared band (10 μm or more):

$$\frac{4\pi}{3}\sum_{i} N_i\alpha_i = \frac{n^2-1}{n^2+2} \tag{3.13}$$

Where n is the refractive index of the medium, $n^2 = \varepsilon$. In the case of isotropic crystals and the field strength of the incident field is weak, χ is a considered as a scalar and P is in linear relationship of E, $P = \chi E$. And in the case of anisotropic crystals or strong incident fields, χ is no longer a scalar, but a tensor above the second order, and the contribution of higher-order terms above the second order must also be considered in the field strength.

3.1.1.3 Classical Light Scattering of Molecular Polarization

For anisotropic molecular systems, the polarizability of molecules is not a scalar but a tensor. The electric dipole moment of the jth molecule can be expressed as:

$$p_i = \sum_{j=x,y,z} \alpha_{ij} E_{Lj} \tag{3.14}$$

In which, α_{ij} is the molecular polarizability tensor, E_{ij} is the local field component, and the electric dipole moment components p_x, p_y, p_z have the following relationship with the local field component E_{Lx}, E_{Ly}, E_{Lz}:

$$p_x = \alpha_{xx}E_{Lx} + \alpha_{xy}E_{Ly} + \alpha_{xz}E_{Lz} \tag{3.15}$$

$$p_y = \alpha_{yx}E_{Lx} + \alpha_{yy}E_{Ly} + \alpha_{yz}E_{Lz} \tag{3.16}$$

$$p_z = \alpha_{zx}E_{Lx} + \alpha_{zy}E_{Ly} + \alpha_{zz}E_{Lz} \tag{3.17}$$

Where tensor α_{ij} can be expressed as:

$$\alpha_{ij} = \begin{pmatrix} \alpha_{xx}\alpha_{xy}\alpha_{xz} \\ \alpha_{yx}\alpha_{yy}\alpha_{yz} \\ \alpha_{zx}\alpha_{zy}\alpha_{zz} \end{pmatrix} \tag{3.18}$$

The polarizability caused by field induction is expanded according to normal coordinates. If only the second-order effect is considered, the expression is:

$$\alpha_{ij} = (\alpha_{ij})_0 + \sum_l \left\{ \left(\frac{\partial \alpha_{ij}}{\partial Q_l} \right)_0 Q_l + \frac{1}{2} \sum_{l,k} \frac{\partial^2 \alpha_{ij}}{\partial Q_l \partial Q_k} Q_l Q_k + \cdots \right\} \tag{3.19}$$

The first term on the right side of equation (3.19) is the zero-order term of the polarizability, and it is a term without frequency shift, which belongs to the elastic scattering Rayleigh scattering of light. The second term is the first derivative term of polarizability with respect to normal coordinates, which corresponds to the first-order Raman scattering. The third term is the second derivative term of the polarizability to the normal coordinate, which corresponds to the second-order Raman scattering. The first-order Raman scattering is taken as an example and discussed by indirect theory.

The incident light field E_i with the frequency of ω_i can be expressed as:

$$E_i = E_0 cos \omega_i t \tag{3.20}$$

Where E_0 is the amplitude of the incident light field. If the incident light is irradiated on the molecule, the induced electric dipole moment p of the molecule is:

$$p = \alpha E_i = \alpha E_0 cos \omega_i t \tag{3.21}$$

Where α is the molecular polarizability. That is the first derivative term of the polarizability with respect to the normal coordinate in equation (3.19). When the molecules are vibrating at an angular frequency ω, the nuclear displacement Q of the molecule can be written as:

$$Q = Q_0 cos \omega t \tag{3.22}$$

Where Q_0 is the amplitude of vibration, and in the case of small amplitude, the molecular polarizability can be expressed as a linear function of the following nuclear displacement Q:

$$\alpha = \alpha_0 + \left(\frac{\partial \alpha}{\partial Q} \right)_0 Q \tag{3.23}$$

In which, α_0 is the polarizability of the nucleus at the equilibrium position, $\left(\frac{\partial \alpha}{\partial Q} \right)_0$ is the change in polarizability caused by the unit nuclear displacement at the equilibrium position. Combining the above three formulas, we can get:

$$\begin{aligned} p &= \alpha E_0 \cos \omega_i t \\ &= \alpha_0 E_0 \cos \omega_i t + \left(\frac{\partial \alpha}{\partial Q} \right)_0 Q_0 E_0 \cos \omega_i t \cdot \cos \omega t \\ &= \alpha_0 E_0 \cos \omega_i t + \left(\frac{\partial \alpha}{\partial Q} \right)_0 Q_0 E_0 \cos (\omega_i + \omega) t \cdot \cos (\omega_i - \omega) t \end{aligned} \tag{3.24}$$

Three frequencies induced by molecular-induced dipole moment are ω_i, $\omega_i - \omega$, and $\omega_i + \omega$. The first term on the right side of equation (3. 24) is based on the frequency of incident light ω_i, which is the same Rayley scattering term as in equation (3.19), and the term of $\omega_i - \omega$ is first-order Stokes Raman scattering; the term of $\omega_i + \omega$ is first-order anti-Stokes Raman scattering.

3.1.2 Light Scattering and Extinction Theory

This section focuses on static light scattering, involving angular scattering mode, total scattering, and absorption. The progress presented here includes all the scattered progress and integrates them in unconventional ways. Mie theory is a general description of the scattering of a single sphere. It is a model developed from a simple scattering model that relies solely on the wave diffraction theory. This model makes the conclusions and physical consciousness of most complex situations organized. To understand the methods of static light scattering more deeply, readers can read some classic books on static light scattering [12–14]. Dynamic light scattering utilizes the time fluctuation of scattered light related to particle diffusion, which is related to particle size. More methods of dynamic light scattering are introduced in [15,16].

3.1.2.1 Differential Scattering Cross-Section

Differential scattering cross-section $dC_{\text{sca}}/d\Omega$ describes the incident light intensity $I_0\left(\text{w/m}^2\right)$ of scattering power P_{sca} at each solid angle unit $\Omega(\text{W/sr})$:

$$\frac{P_{\text{sca}}}{\Omega} = \frac{dC_{\text{sca}}}{d\Omega} I_0 \tag{3.25}$$

Therefore, the unit of $dC_{\text{sca}}/d\Omega$ is m^2 / sr. The incident light intensity refers to the incident power per unit area in the incident direction.

The incident light intensity refers to the incident power per unit area in the incident direction.

$$I_{\text{sca}} = P_{\text{sca}}/A \tag{3.26}$$

The solid angle of the detector with a distance of r from the scatter is:

$$\Omega = \frac{A}{r^2} \tag{3.27}$$

Therefore, it can be obtained from the above two formulas:

$$I_{\text{sca}} = I_0 \frac{dC_{\text{sca}}}{d\Omega} \times \frac{1}{r^2} \tag{3.28}$$

This equation is known as $1/r^2$, 3D geometry based on space. Note that the scattered light intensity is directly proportional to the differential cross-section. Therefore, when referring to these two concepts, the meaning is the same.

3.1.2.2 Efficiency

The scattering or absorption efficiency Q is the ratio of any full scattering cross-section to the projection of the incident direction of the scatterer plane, which is a dimensionless parameter:

$$Q_{\text{sca or abs}} = \frac{C_{\text{sca or abs}}}{A_{\text{proj}}} \quad (3.29)$$

For a sphere with radius a:

$$A_{\text{proj}} = \pi a^2 \quad (3.30)$$

The efficiency is physically intuitive because they compare the optical cross-section with the geometric cross-section. If the light has no fluctuation, that is, it is merely a single particle, the sum of its scattering and absorption efficiency is independent of the particle size.

3.1.2.3 Extinction and Reflectivity

The attenuation of light intensity after passing through the particle system is called extinction. When the light passes through the medium, the light intensity attenuates in an exponential form, and the attenuation law follows Lambert Beer's law:

$$I_{\text{trans}} = I_0 e^{-\tau x} \quad (3.31)$$

Here, τ is the turbidity of the medium, which is related to the number concentration n of particles and its extinction section C_{ext}:

$$\tau = nC_{\text{ext}} \quad (3.32)$$

Extinction is caused by scattering and absorption, where scattering deviates from the direction of the incident path, and absorption converts light into other forms of energy (such as heat).

$$C_{\text{ext}} = C_{\text{abs}} + C_{\text{sca}} \quad (3.33)$$

Reflectivity ω is the ratio of scattering to extinction:

$$\omega = C_{\text{sca}} / C_{\text{ext}} \quad (3.34)$$

A related parameter is the Rayleigh ratio. For a particle system, the Rayleigh ratio is equal to the scattering cross-section. Both the differential Rayleigh ratio and the total Rayleigh ratio are multiples of the particle number concentration. Therefore, the unit of the Rayleigh ratio can be either $(\text{m} \cdot \text{sr})^{-1}$ or m^{-1}. Rayleigh ratios are often used to describe light scattering and attenuation in gases and liquids [16].

3.1.3 Classification of Light Scattering

3.1.3.1 Small Particles: Rayleigh Scattering

If all the sizes of small particles are smaller than the wavelength of light, the scattering on the small particles is called Rayleigh scattering. In Rayleigh scattering, the length function has two dimensions, which can be expressed by a simple dimension parameter. Since the particle size is much smaller than the wavelength of light, the phase of incident light passing through the entire small particle volume is uniform. If the particle is subdivided into infinitesimal sub volumes, all sub volumes have the same phase. In addition, since the particles are relatively small, the distance between the light scattering from the sub volume of all particles and the detector is actually equal. That is to say, there is no accumulation of relative phases in the propagation process. Because of this, the light emitted by each sub volume of particles has the same phase on the detector. The light intensity is the square of the field amplitude, so:

$$I \propto V_{\text{part}}^2 \tag{3.35}$$

The cross-section is an effective area, so its unit is the square of the length. So far, V_{part} in equation (3.35) is the sixth power of length. Therefore, the factor of the inverse fourth power of the length is missing. There are only two length scales in this problem: particle size and light wavelength. Particle size has been used in V_{part}. Therefore, in order to make the cross-section have a suitable unit, it is necessary to include a factor to obtain a cross-section with a suitable unit. Therefore:

$$I \propto \lambda^{-4} V_{\text{part}}^2 \tag{3.36}$$

The simple part derived does not depend on the shape of the particle, so the result is independent of the shape. This part is also independent of the scattering angle, so the scattering is independent of the scattering angle, which means the scattering is isotropic.

For the simplest case of spherical particles with radiusa, the electromagnetic theory can be accurately expressed by Rayleigh scattering theory, which can be defined as the particle size parameter:

$$a = \frac{2\pi a}{\lambda} \tag{3.37}$$

This formula contains the dimensionless ratio of two length scales. The conditions of Rayleigh scattering include:

$$\begin{gathered} a \ll 1 \\ ma \ll 1 \end{gathered} \tag{3.38}$$

Where m is the relative index of the particle refraction angle:

$$m = n_{\text{particulate}} / n_{\text{medium}} \tag{3.39}$$

The differential cross-section is:

$$\frac{dC_{\text{sca}}}{d\Omega} = k^4 a^6 \left|\frac{m^2-1}{m^2+2}\right|^2 = \frac{16\pi^4 a^6}{\lambda^4}\left|\frac{m^2-1}{m^2+2}\right|^2 \tag{3.40}$$

If inserted in equation (3.28), we get:

$$I_{vv} = \frac{k^4 a^6}{r^2}\left|\frac{m^2-1}{m^2+2}\right|^2 I_0 \tag{3.41}$$

If the incident light is non-polarized, this expression is rewritten as:

$$I = \frac{k^4 a^6}{2r^2}\left|\frac{m^2-1}{m^2+2}\right|^2 (1+\cos\theta) I_0 \tag{3.42}$$

Here, $k = 2\pi/\lambda$, it usually refers to the refractive index term and Lorentz, which can be abbreviated as:

$$F(m) = \left|\frac{m^2-1}{m^2+2}\right|^2 \tag{3.43}$$

Which could simplify equation (3.40) as:

$$\frac{dC_{\text{sca}}}{d\Omega} = k^4 a^6 F(m) \tag{3.44}$$

Rayleigh scattering has some important characteristics:

1. Homogeneity and homogeneity. I_{vv} is independent of angle θ of the scattering plane. $I_{\text{VH}} = I_{\text{HV}} = 0$ and $I_{\text{HH}} = I_{vv}\cos^2\theta$.
2. Correlation of λ^{-4}. Blue light scatters more than red light, which is related to the blue of the blue sky and the red of the sunset [17,18] but also includes other (or less) factors. Under excellent clean air conditions (no particles), molecular scattering occurs due to small thermodynamic fluctuations in the air. Since this fluctuation is smaller than the wavelength, Rayleigh λ^{-4} is related to the fluctuation.
3. Tyndall effect. The particle size strongly depends on $V_{\text{part}}^2 \sim a^6$, which leads to the increase of scattering in the coarse particle system. To understand this, assume that the Rayleigh scattering of n particles in unit volume is proportional to the total scattering in the particle system:

$$I_{\text{sca}} \propto nV_{\text{part}}^2 \tag{3.45}$$

If the only growth process in the system is condensation, the total particle mass is conserved. Therefore, nV_{part} is always unchanged. On the other hand, V_{part} increases due to condensation. Equation (3.45) is rewritten as:

$$I_{\text{sca}} \propto nV_{\text{part}} \cdot V_{\text{part}} \tag{3.46}$$

This shows that with the aggregation of the system, the scattering intensity increases in proportion to V_{part}, which is Tyndall effect. Pay attention to the reverse Tyndall effect. For a certain amount of material, the finer it is divided, the less it scatters.

The set of all 4π solid angle differential sections produce the full section. In the scattering arrangement, the incident light is polarized in the vertical direction. It can be deduced that:

$$C_{sca} = \int \frac{dC_{sca}}{d\Omega} d\Omega = \frac{dC_{sca}}{d\Omega} \int_0^{2\pi} \int_{-1}^{1} \left(1 - \cos^2 \varphi \sin^2 \theta\right) d(\cos\theta)\, d\varphi = \frac{8\pi}{3} \frac{dC_{sca}}{d\Omega} \tag{3.47}$$

Since the differential cross-section is independent of angle, it can be seen that the factor $8\pi / 3$ comes from the integral of polarization. From equations (3.40) and (3.47), we can get:

$$C_{sca} = \frac{8\pi}{3} k^4 a^6 F(m) \tag{3.48}$$

Therefore, the scattering efficiency $Q = C_{sca} / \left(\pi a^2\right)$ is:

$$Q_{sca} = \frac{8}{3} a^4 F(m) \tag{3.49}$$

For Rayleigh scattering, the size parameter $a \ll 1$. Therefore, equation (3.49) shows that Rayleigh scatterers are not particularly effective, that is, their scattering is much smaller than that represented by the geometric cross-section.

Rayleigh absorption cross-section is expressed by the following formula:

$$C_{abs} = \frac{8\pi^2 a^3}{\lambda} \operatorname{Im}\left(\frac{m^2 - 1}{m^2 + 2}\right) \tag{3.50}$$

In which, Im is the imaginary part. By applying $m = n + ik$ and $\operatorname{Im}\left[\left(m^2 - 1\right) / \left(m^2 + 2\right)\right] = E(m)$, the absorption efficiency can be simplified as:

$$Q_{abs} = 4aE(m) \tag{3.51}$$

As for scattering, a simple dimensional demonstration of the absorption cross-section can be made. Because the particles are very small, the light wave can completely penetrate the particles. Therefore, the absorption amount of the whole sub volume of the particles is equal to the application of $C_{abs} \propto V_{part}$. In order to match the unit with the unique length scale divided in the system, $C_{abs} \propto \lambda^{-1} V_{part}$ is obtained, which is completely different from scattering.

Extinction is the sum of scattering and absorption. If the refractive index of a particle is completely real and there is no absorption, extinction is equal to scattering. If there is an imaginary part of the particle's refractive index, the absorption in the Rayleigh rule will dominate the scattering. Here, the size parameter is very small, so there is an additional factor for scattering a^3 [comparing equations (3.48) and (3.50)], which is much smaller than the whole Rayleigh particle.

These facts make it possible to measure the particle size and number concentration of small particles by light scattering. The scattering light intensity at any angle is proportional to the cross-section and particle number concentration, so $I_{\text{sca}} \sim na^6 F(m)$ can be obtained from equation (3.44).

Usually, these measurements can be made by calibrating a known scatterer, such as a gas or liquid, with a known Rayleigh proportional coefficient. The turbidity is measured synchronously according to equation (3.31). It can be seen from equation (3.32) that the turbidity is related to the number concentration and extinction cross-section. It is defined here that the extinction cross-section is equal to the absorption cross-section. Then by applying equation (3.46) and $\tau \sim na^3 E(m)$, the two unknowns n and a can be determined by two formulas.

3.1.3.2 Soft Particles: Rayleigh–Debye–Gans scattering

If the refractive index difference between the scattering particle and the medium is small, that is to say, m is very close to unity, then the interesting situation of Rayleigh–Debye–Gans scattering will occur. The scattering conditions of Rayleigh–Debye–Gans scattering are as follows:

$$|m-1| \ll 1 \tag{3.52}$$

$$\rho = 2a|m-1| \ll 1 \tag{3.53}$$

Note that equations (3.52) and (3.53) are applicable to "soft" particles of any size, $m \to 1$. The parameter ρ is called the phase shift parameter, which represents the phase difference of the wave when the light wave propagates through the particle directly across the particle diameter and across the same distance in the medium.

We note that for X-rays, the refractive index is nearly uniform, so the Rayleigh–Debye–Gans limit is applicable to this radiation. Indeed, most of the relevant scenarios developed later are to try to understand and describe small angle X-ray scattering (SAXS) and small angle neutron scattering (SANS). Here, this important correlation is applied as the basis for understanding the light scattering of particles with arbitrary refractive index.

For a sphere with radius a, the Rayleigh–Debye–Gans differential scattering cross-section on the vertical polarized light scattering plane is:

$$\frac{dC_{\text{sca}}}{d\Omega_{\text{RDC}}} = \frac{dC_{\text{sca}}}{d\Omega_{\text{R}}}\left[\frac{3}{u^3}(\sin u - u\cos u)\right]^2 \tag{3.54}$$

In which,

$$u = 2\alpha \sin(\theta/2) \tag{3.55}$$

or

$$u = qa \tag{3.56}$$

In equation (3.54), subscripts RDG and R represent Rayleigh–Debye–Gans scattering and Rayleigh scattering, respectively [2]. In equation (3.56), q is the scattered wave vector, which is given in the following formula:

$$q = (4\pi / \lambda)\sin(\theta / 2) \tag{3.57}$$

Scattered wave vector q is a very important variable, which has the same pedigree as SAXS and SANS. The unit of q is the reciprocal of the unit of length, so q^{-1} is the length scale in scattering experiments. This means that when the experimenter adjusts the scattering angle to determine q, the scattering in the scattering system is at the perceived length q^{-1}. Applying this concept to the scattering of particles with arbitrary radius a means that if q changes, but qa is always less than the unit quantity, there will be no dependence on q, that is, it has nothing to do with the scattering angle. This happens when the scattering length value $q^{-1} > a$, so the sample cannot be resolved in this case. This is the isotropic forward scattering law of particles of any size. Only when $q^{-1} < a$, the scattering can analyze the sample, and the scattered light intensity shows the dependence on q at an angle. Moreover, the scattering starts from $qa \approx 1$, and this relationship can be used to determine the particle size.

Rayleigh–Debye–Gans scattering is the diffraction limit of scattering; When $m \to L$, the electromagnetic properties are suppressed. The term in parentheses of equation (3.54) is the Fourier transform of the sphere, so it represents the diffraction of the sphere. This is completely similar to the famous single-slit Fraunhofer diffraction. Therefore, RDG is a simple Fourier transform of sphere and surface area. In quantum mechanics, this similarity is the first approximation with weak scattering potential.

Rayleigh–Debye–Gans scattering is not limited to spheres. Similar to equation (3.54), the scattering of particles of any shape is the square of Fourier transform times the Rayleigh scattering cross-sectional area.

In the scattering diagram of RDG, when $qa \leq 1$, the scattered light intensity is always consistent with q, so it is consistent with the scattering angle, which is called "forward scattering lobe". The forward scattering lobe starts from $qa = 1$, which is equivalent to a small angle $\theta = \lambda / (2\pi a)$, i.e., $\sin\theta / 2 \approx \theta / 2$. Here, the scattering magnitude is equal to the Rayleigh result in equation (3.40).

3.1.3.3 Sphere of Arbitrary Size and Refractive Index: Mie Scattering Theory

Rayleigh theory and Rayleigh–Debye–Gans theory about scattering and absorption are representative theories for solving Maxwell's equations. Due to their small size and refractive index, approximate solutions are applied to the equations. For any particle, the solution of Maxwell's equation must be accurate. Mie first applied this solution to the simplest case of a homogeneous sphere, and the concept of "meter scattering" is often applied to this case. However, this equation is not particularly simple to apply or cannot simply get physical meaning. Based on the experience of RDG scattering, the scattering wave vector q (dimensionless qa is better) is used to draw the Mie scattering differential interface. Then, its physical model [19,20] provides a normative description method for arbitrary sphere scattering.

In the previous section, we briefly introduced the elastic scattering of light (Rayleigh scattering and Mie scattering). If the size of the scattering particle is smaller than the wavelength of the incident light, it falls into the scope of Rayleigh scattering. If the size of the scattering particle is equal to or larger than the wavelength of the incident light, it belongs to Mie scattering. We can simply express these two kinds of scattering criteria as follows: Rayleigh scattering satisfies $\alpha << 1$, and the Mie scattering satisfies $|m|\alpha \ll 1$, where α is a dimensionless parameter:

$$\alpha = \frac{2\pi a}{\lambda} \tag{3.58}$$

$$\lambda = \frac{\lambda_0}{m_0} \tag{3.59}$$

In which, α is the radius of the scattering particle, 2 is the relative scattering wavelength, λ_0 is the vacuum incident wavelength, m is the refractive index of the dielectric ball, and m_0 is the refractive index of the medium around the ball.

When studying the Mie scattering of a single particle, we consider the particle as a spherical particle. Figure 3.1 shows the geometric configuration of Mie scattering of a dielectric ball [11]. The refractive index of the ball is m, and the scattering angles are ϕ and θ. For the case where the surrounding medium is air, $m_0 = 1$, as long as the refractive index of the material constituting the particle is known, we can calculate it. If the material constituting the particles has coordinated chemical components, the

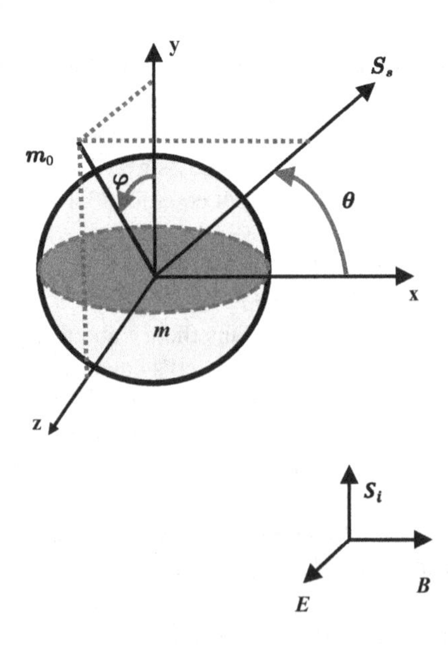

FIGURE 3.1 Geometric configuration of Mie scattering of dielectric spheres.

effective refractive index must be taken into account in the calculation at a certain wavelength. The refractive index m of the dielectric sphere can be expressed as the following complex number:

$$m = n + i\kappa \tag{3.60}$$

Here, n is the refractive index of light, which is the speed of light in a vacuum divided by the speed of light in the medium. The order of magnitude of the complex refractive index is $|m| = \left(n^2 + \kappa^2\right)^{1/2}$. The imaginary part of the complex number indicates that it is related to absorption. They have the following relationship:

$$\alpha = \frac{\pi\kappa}{\lambda} \tag{3.61}$$

Relative to the scattering plane determined by the incident light and the scattered light, the vertical and horizontal components of the scattered radiation can be expressed as:

$$I_\Phi = I_0 \frac{\lambda\wedge 2}{4\pi^2 r^2} I_1 \sin^2\phi \tag{3.62}$$

$$I_\theta = I_0 \frac{\lambda\wedge 2}{4\pi^2 r^2} I_2 \cos^2\phi \tag{3.63}$$

Where I_0 is the incident light intensity and I have the following relationship:

$$I_1 = \left|\sum_{n=1}^{\infty} \frac{2n+1}{n(n+1)}\left[a_n\pi_n(\cos\theta) + b_n\tau_n(\cos\theta)\right]\right|^2 \tag{3.64}$$

$$I_2 = \left|\sum_{n=1}^{\infty} \frac{2n+1}{n(n+1)}\left[a_n\pi_n(\cos\theta) + b_n\tau_n(\cos\theta)\right]\right|^2 \tag{3.65}$$

The functions related to the scattering angle in equations (3.64) and (3.65) are not Legendre polynomials

$$\pi_n(\cos\theta) = \frac{P_n^{(1)}(\cos\theta)}{\sin\theta} \tag{3.66}$$

$$\tau_n(\cos\tau) = \frac{dP_n^{(1)}(\cos\theta)}{d\theta} \tag{3.67}$$

$$a_n = \frac{\psi_n(\alpha)\psi_n'(m\alpha) - m\psi_n(m\alpha)\psi_n'(\alpha)}{\xi(\alpha)\psi_n'(m\alpha) - m\psi_n(m\alpha)\xi_n'(\alpha)} \tag{3.68}$$

$$b_n = \frac{m\psi_n(\alpha)\psi_n'(m\alpha) - \psi_n(m\alpha)\psi_n'(\alpha)}{m\xi(\alpha)\psi_n'(m\alpha) - \psi_n(m\alpha)\xi_n'(\alpha)} \tag{3.69}$$

Dimensional parameter is:

$$\alpha = \frac{2\pi a m_0}{\lambda_0} \tag{3.70}$$

The Ricatti–Bessel function ψ and ξ can be expressed by the semi-integer Bessel function of the first kind $J_{n+\frac{1}{2}}(z)$:

$$\psi_n(z) = \left(\frac{\pi z}{2}\right)^2 J_{n+\frac{1}{2}}(z) \tag{3.71}$$

$$\xi_n(z) = \left(\frac{\pi z}{2}\right)^{\frac{1}{2}} H_{n-\frac{1}{2}}(z) = \psi_n(z) + iX_n(z) \tag{3.72}$$

Here $H_{n-\frac{1}{2}}(z)$ is the second type of semi-integer Hankel function, and the parameter x is determined by the second type of $Y_{n+\frac{1}{2}}(z)$ semi-integer Hankel function, that is:

$$X_n(z) = -\left(\frac{\pi z}{2}\right)^{\frac{1}{2}} Y_{n+\frac{1}{2}}(z) \tag{3.73}$$

The differential scattering components related to polarization are:

$$\sigma_{\text{VV}} = \frac{\lambda^2}{4\pi^2} I_1 \tag{3.74}$$

$$\sigma_{\text{HH}} = \frac{\lambda^2}{4\pi^2} I_2 \tag{3.75}$$

The subscript VV of the scattering cross-section indicates the polarization direction of the incident light relative to the scattering plane and the polarization direction of the scattered light along the vertical direction, and the subscript HH indicates the polarization direction of the incident light relative to the scattering plane and the polarization direction of the scattered light along the horizontal direction. For the case where the human light is not polarized, the total scattering cross-section can be obtained from equations (3.64), (3.65), (3.74), and (3.75):

$$\sigma_T = \frac{\lambda^2}{8\pi^2}(I_1 + I_2) \tag{3.76}$$

Thereby obtaining a component of the scattering intensity:

$$I_{\text{vv}} = I_0 \frac{1}{r^2} \sigma_{\text{VV}} \sin^2\phi \tag{3.77}$$

$$I_{\mathrm{HH}} = I_0 \frac{1}{r^2} \sigma_{\mathrm{HH}} \cos^2 \phi \tag{3.78}$$

$$I_T = I_0 \frac{1}{r^2} \sigma_T \tag{3.79}$$

3.2 PARTICLE SIZE INVERSION BASED ON LIGHT SCATTERING

The total light scattering particle size measurement method is to obtain the particle size distribution of the particle system by measuring the extinction values at multiple wavelengths. In the data processing of this measurement method, we will encounter the problem of solving the first kind of Fredholm integral equation, so we need to determine the upper and lower limits of the integral equation in advance, which is the particle size measurement range. If there is a large gap between the selected particle size range and the actual particle size range, incorrect inversion results may be obtained. The particle size measurement range of the spherical particle system can be expressed by the minimum and maximum particle diameters. According to the principle of equivalent extinction, the particle size measurement range of non-spherical particle system can be approximately equivalent to the particle size measurement range of spherical particle system with equal volume. On the basis of expounding the principle of total light scattering method for spherical particle size measurement, this chapter deeply studies the particle size measurement range of typical spherical particle systems in the visible wavebands and visible infrared wavebands, so as to provide a reference for determining the particle size measurement range of different particle systems in different wavebands.

3.2.1 Research on Independent Mode Inversion Algorithm

When the particle size, relative refractive index, and wavelength are given, the particles will scatter under this light. It is easy to solve the spatial distribution of the scattering field according to Mie scattering theory, which is called a positive problem. On the contrary, using the distribution of incident light and scattering field, the particle size and particle size distribution of its scattering particles are solved, which is called the inverse problem. Particle size inversion algorithm includes independent and non-independent models. The difference between the two types of algorithms is whether the particle size distribution function needs to be assumed in advance.

At present, due to the widespread application in many fields, the research on inverse problems has made great progress, and also promoted the progress of inversion algorithms to calculate laser particle size. The commonly used algorithms include the least mean square algorithm, inversion algorithm, projection inversion algorithm and Chahine inverse algorithm. The calculation process of the independent mode inversion algorithm is to assume the initial value first, and then calculate according to the specific iteration format. A solution value is obtained after some calculations, and the modified solution value is very close to the true value. However, the inversion algorithm is not stable enough, and its convergence characteristics are not good. In general, the algorithm is greatly affected by noise, which will lead to inaccurate solution values.

3.2.2 Research on Dependent Inversion Algorithm

The dependent inversion algorithm is called the function limitation method. This algorithm requires the knowledge of the particle size information of the particle system to be measured, which usually means that the particle size distribution in accordance with a certain functional relationship needs to be known and solved based on this model. In most cases, the particle size distribution of industrial commonly used grinding particles conforms to some kind of double parameter or even multi-parameter function, including logarithmic normal distribution, Rosin Rammler distribution, and normal distribution. The specific calculation process is as follows: firstly, according to the function (usually the double parameter distribution function) relationship that this algorithm conforms to, and under the known multi-wavelength, to calculate the extinction value of the particle system (the more, the better); secondly, to compare the calculated value with the measured value, and use the optimization inversion algorithm to calculate their ratio or square difference to find the distribution function that makes the value the minimum.

Fine particles under natural conditions are composed of suspended particles with different particle sizes. Suppose that in the medium of unit volume, the number of particles with particle size between D and $D+dD$ is $f\ (D)\cdot dD$, and the particle size distribution function of suspended particles is $f\ (D)$. The scattering light intensity with the wavelength of λ, and space angle θ in the direction is:

$$I(\theta,\lambda,\varphi)=\frac{\lambda_j^2 I_0}{4\pi^2 r^2}\int_{D_{\sin}}^{D_{\max}}\left[i_1(\theta)\sin^2\varphi+i_1(\theta)\sin^2\varphi\right]\cdot f(D)dD \tag{3.80}$$

Where $D_{\min}$ and $D_{\max}$ represent the upper and lower limits of particle size in the medium, respectively.

In the measurement of particle size distribution of fine particles, the scattering light intensity values at different angles or wavelengths are generally measured first, then the particle size distribution function $f(D)$ of the number of suspended particles is inversed according to equation (3.80). It is actually Fredholm's first type integral equation, which is a typical ill-posed equation. When the value $I(\theta,\lambda,\varphi)$ is slightly changed, the particle number particle size distribution function $f(D)$ solved according to will change greatly. The discretization of equation (3.80) is expressed as:

$$I(\theta,\lambda_j,\varphi)=\frac{\lambda_j^2 I_0}{4\pi^2 r^2}\sum_{h=1}^{m}\left[i_1(\theta)\sin^2\varphi+i_1(\theta)\sin^2\varphi\right]\cdot f(D_h) \tag{3.81}$$

The subscribe h is the discrete number of particle size distribution function $f(D)$, that is, the particle size is divided into M grades, j is the number of different wavelengths selected in the spectral measurement, with a total of S; $I(\theta,\lambda,\varphi)$ is the measured value, and I_0 is the incident light intensity. Assume $C=\frac{1}{4\pi^2 r^2}$ be a constant term, then:

$$\frac{I(\theta,\lambda_j,\varphi)}{I_0}=C\lambda_j^2\sum_{h=1}^{n}\left[i_1(\theta)\sin^2\varphi+i_1(\theta)\sin^2\varphi\right]\cdot f(D_h) \tag{3.82}$$

Assign $g=\left[\frac{I(\theta,\lambda_1,\phi)}{I_0},\frac{I(\theta,\lambda_2,\phi)}{I_0},\ldots,\frac{I(\theta,\lambda_s,\phi)}{I_0}\right]^T$, and the elements in matrix T are: $t_{j,h}=C\lambda_j^2\left[i_1(\theta)\sin^2\varphi+i_1(\theta)\sin^2\varphi\right](j=1,2,\ldots,S;h=1,2,\ldots,M)$; the particle size distribution function is $f=\left[f(D_1),f(D_2),\ldots f(D_u)\right]^T$. Then equation (3.82) can be simplified as $g=Tf$, and $f=T^{-1}g$, Given column vector g and coefficient matrix T, particle size distribution f can be determined by solving equations.

The elements t in T are known to be a coefficient matrix independent of particle size distribution. If the optical parameters of the particle to be measured are known, the matrix can be calculated using classical Mie theory. However, from the calculation results, the T-matrix is a seriously ill conditioned matrix, and its row modulus condition number is more than 200, which is far more than 1. Therefore, when the matrix inversion or other methods are directly used, a reasonable and effective result cannot be obtained. Here, an independent algorithm is proposed to solve the T-matrix, and then f can be calculated.

3.3 FINE PARTICLE MEASUREMENT METHOD

The measurement method of fine particles is based on the above light scattering theory. When there are particles in the medium, laser irradiation on the surface of fine particles will produce scattered light, and the parameters of scattered light are closely related to the particle size. The photoelectric conversion element is used to receive the scattered light signal in a certain direction. When each particle passes through the laser area, it will cause a pulse signal. Therefore, the number of particles can be monitored by counting the number of pulse signals in a unit of time. Focusing on the amplitude of the pulse signal, the particle size of the measured particle can be calculated according to the light scattering theory. The particle measurement equipment thus developed is called optical particle counter, which has the advantages of high sensitivity, less impact from external noises, wide particle size measurement range, and is suitable for online measurement [21].

3.3.1 Operating Principle of Optical Particle Counter

Laser irradiation on the particle surface will generate scattered light in the entire space angles. The scattered light can be divided into two categories according to the signal acquisition mode, namely static light scattering method and dynamic light scattering method.

3.3.1.1 Static Light Scattering Method

When a beam of light shines on the particles, it will cause the scattering of light on the particles. The frequency shift, angular distribution, degree of polarization, and illumination of the scattered light are related to the particle size, shape, and interaction between particles. Each scattering element in the scattering volume scatters the incident light. The distribution of scattering light intensity with scattering angle contains the information of particle size. The corresponding particle size distribution can be obtained by measuring the angular distribution of light intensity or the distribution of light energy with solid angle. This method measures the average time value of particle scattering light intensity, namely static light scattering method.

According to the spatial position of scattered light detection, the static light scattering method can be divided into total light scattering method (transmission method), forward small angle scattering method, vertical scattering method, and backward small angle scattering method. Each of these methods is described below.

3.3.1.1.1 Total Light Scattering Method

The total light scattering method refers to the transmission method, also known as the extinction method, which is a special scattering method [22,23]. When the light beam reflected from the light source passes through the particles, due to the scattering effect of particles, the light intensity of the transmitted light is less than that of the incident light, and its attenuation is related to the particle size. The measurement principle of total light scattering method is shown in Figure 3.2. A laser parallel beam converges on the particle surface through the lens, and the transmitted light through the particle converges through two lenses and is detected by the laser diode.

Total light scattering particle measurement technology is a relatively simple technology in terms of both measuring principle and measuring device, and its data acquisition and processing are relatively simple. Therefore, the total scattering method technology has received extensive attention and has successively developed several other methods to measure particle concentration and particle size distribution at the same time, such as single wavelength method, dual wavelength method, multi-pair wavelength method, extinction voltage generation method, and the photo resistance method, as well as the multi-wavelength method. The development of these measurement methods provides a basis for the online and real-time application of light-scattering particle measurement technology. The disadvantage of this method is that when measuring a single particle, most of the light is transmitted directly, resulting in almost constant transmitted light intensity, which requires very high resolution and stability of photoelectric receiving elements and amplifiers.

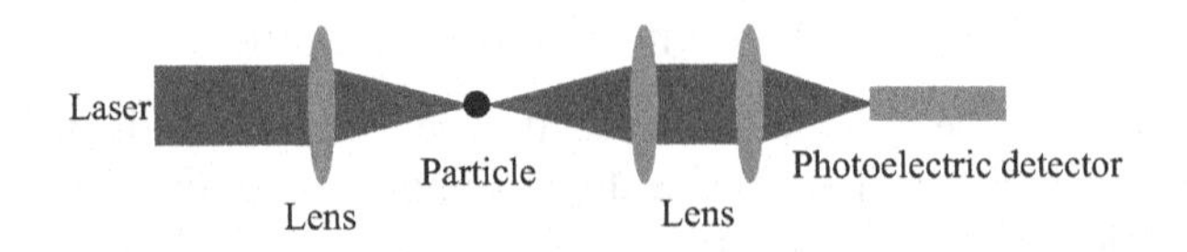

FIGURE 3.2 Principle of particle measurement by total light scattering method.

3.3.1.1.2 Forward Small Angle Scattering Method

Figure 3.3 shows the spatial distribution of scattering light intensity with the incident light wavelength 780 nm, the particle refractive index 1.44, and the particle diameters 0.3, 1, and 3 μm. Obviously, the scattering light near the transmission light direction is much stronger than other angles. The scattering light with an angle less than 90° with the transmission light is defined as forward scattering light. Due to the high intensity of forward scattered light, it has the advantages of higher particle size resolution and lower limit of particle size detection.

The principle of forward small angle scattering particle measurement is shown in Figure 3.4. Photodiodes are used to measure the scattered light energy distribution of particles in a small forward angle range, from which the number and size of particles can be measured. For larger particles, the small angle forward scattering method is also called diffraction scattering method because the scattering in the small angle forward range is mainly diffraction. The upper limit of this method can

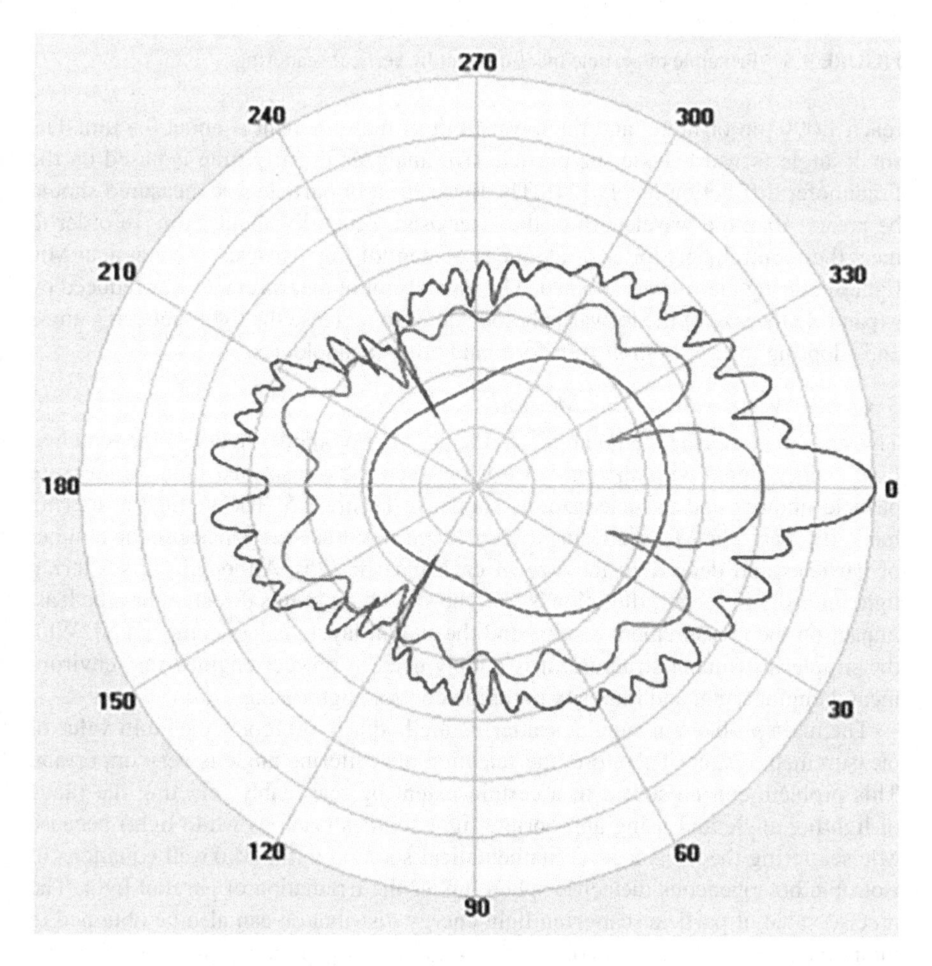

FIGURE 3.3 Spatial distribution of scattered light intensity.

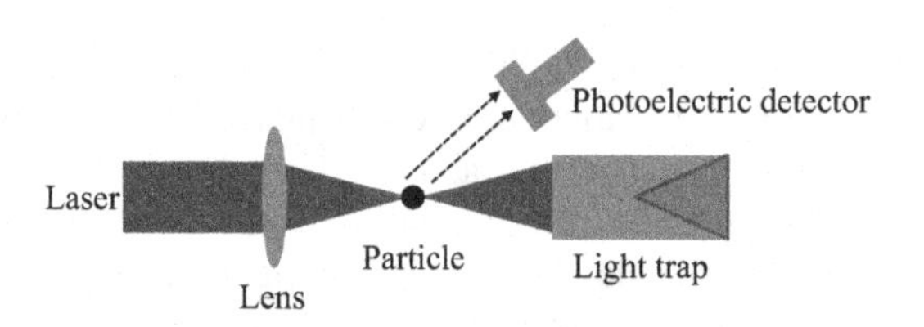

FIGURE 3.4 Principle of particle measurement by forward small angle scattering.

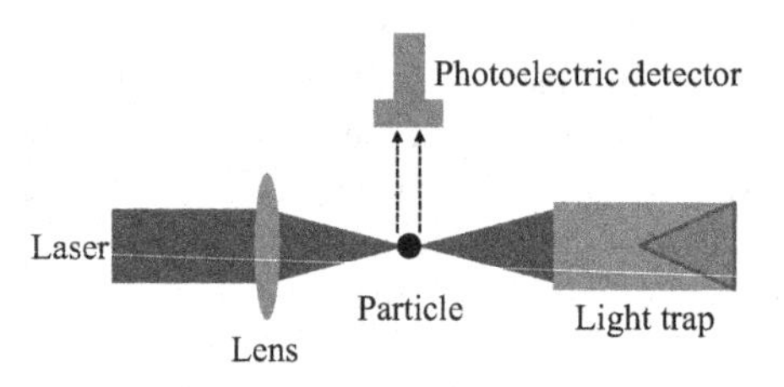

FIGURE 3.5 Principle of particle measurement by vertical scattering.

reach 1,000 μm or more, and the lower limit of measurement is about 0.5 μm. The small angle forward scattering particle size analyzer in early time is based on the Fraunhofer diffraction theory [24]. The lower limit of particle size measured should be greater than the wavelength of the laser used, generally about 2 μm. In order to meet the requirements for accurate measurement of fine particles, the classical Mie light scattering theory was adopted. The lower limit of measurement was reduced by expanding the size of the forward photodetector to increase the light scattering angle and adopting inverse Fourier transform and other technologies.

3.3.1.1.3 Vertical Scattering Method

The vertical scattering method refers to the use of photodiodes to detect the scattered light energy signal with an angle of 90° between the transmitted light, to measure particle number and particle size, as shown in Figure 3.5. The vertical scattering method is based on Mie scattering theory. It can simultaneously measure the number of particles and determine the size of each particle [25]. Although the scattering light intensity in the 90° direction is low, the stray light in this direction has the least impact on the measurement results, and the spatial layout is less complicated. With the simple instrument structure, it is widely used in powder engineering, environmental engineering, and impurity pollution control engineering.

The main problem in angular scattering method is to overcome the multi-value of measurement results. Therefore, the selection of scattering angle is very important. This problem can be solved to a certain extent by reasonably selecting the range of lighting angle and using appropriate light sources (such as white light) because Mie scattering theory is a strict mathematical solution to the Maxwell equations of isotropic homogeneous dielectric sphere under the irradiation of parallel light. The precise value of particle scattering light energy distribution can also be obtained in the range of small particles, so this method has high measurement accuracy and a wide measurement range.

3.3.1.1.4 Small Angle Backscattering Method

The principle of small angle backscattering particle measurement is shown in Figure 3.6. Photodiodes are used to measure the light intensity of scattered light in the direction with an angle less than 90° from the incident light, from which the number and size of particles can be measured. Due to the weak backscattering light intensity of particles, backscattering is seldom used in particle detection. Backscattering coefficient is an important intrinsic optical quantity to characterize the characteristics of water bodies and an important method to study the concentration and size distribution of water particles.

3.3.1.2 Dynamic Light Scattering Method

For a large number of fine particles in the liquid that do irregular motion (Brownian motion), the scattered light of each particle in a certain direction can have different phases. The phase of the scattered light is related to the position of the scattered light particles and the interaction between the scattered particles. Due to the thermal movement of particles, the position of scattering particles changes, resulting in fluctuations in the total light intensity of scattering. In the frequency domain, the scattering light generated by fine particles has a small frequency shift relative to the incident light. By measuring the fluctuation of scattering light or the small frequency shift of scattering light, the relevant dynamic parameters of particle size (translational diffusion coefficient and rotational diffusion coefficient) can be obtained. Then, according to Stokes–Einstein relation, the shape and size parameters of particles can be obtained, which is called dynamic light scattering method.

Figure 3.7 shows the principle diagram of dynamic light scattering particle measurement. The laser emitted by the laser passes through the polarizer and then shines on the particles in the sample cell through the incident lens. The scattered light at a

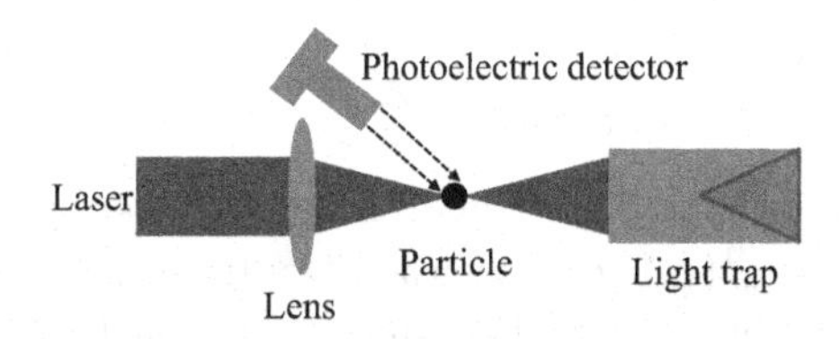

FIGURE 3.6 Particle measurement principle by small angle backscattering.

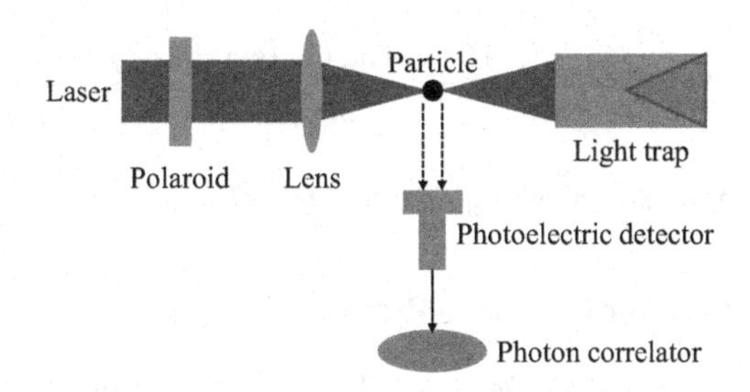

FIGURE 3.7 Principle of particle measurement by dynamic light scattering.

certain angle of the particles enters the photodetector through the receiving light path. The autocorrelation function of light intensity is measured by the photon correlator:

$$G_\theta^{(2)}(\tau_j) = \lim_{M\to\infty} \sum_{k=1}^{M} i(\tau_k)\cdot i(\tau_{k+j})/M \tag{3.83}$$

Where $G_\theta^{(2)}(\tau_j)$ is the autocorrelation function of light intensity, $i(\tau_k)$ represents scattered light intensity signal at time τ_k, τ_j is the delay time, and M is the total number of photons. The electric field autocorrelation function and the light intensity autocorrelation function obey the Siegert relationship [26], which is expressed as:

$$G_\theta^{(2)}(\tau_j) = B\left(1+\beta\left|g_\theta^{(1)}(\tau_j)\right|^2\right) \tag{3.84}$$

Where B is the experimental baseline, $\beta(\le 1)$ is the spatial coherence factor, $g_\theta^{(1)}(\tau_j)$ is the normalized electric field autocorrelation function, $j(1\le j\le M)$ represents the jth channel of the photon correlator.

Electric field correlation function $g_\theta^{(1)}(\tau_j)$ at scattering angle θ is:

$$g_\theta^{(1)}(\tau_j) = \sum_{i=1}^{N} \exp\left(-\frac{16\pi k_B T n_m^2 \sin^2(\theta/2)}{3\eta d_i \lambda_0^2}\cdot\tau_j\right) f(d_i) \tag{3.85}$$

Wherein k_B, T, n_m, η, d, λ_0 are Boltzmann constant, absolute temperature of sample solution, refractive index of solution, viscosity coefficient of solution, particle size and wavelength of incident light in vacuum, respectively; $f(d_i)$ is discrete particle size distribution, with N discrete points in total. Equation (3.4.3) can be simplified as:

$$\boldsymbol{g}_\theta^{(1)} = A_\theta \boldsymbol{f} \tag{3.86}$$

Where $g_\theta^{(1)}$ is a vector composed of normalized electric field autocorrelation function data, with the element $g_\theta^{(1)}(\tau_j)$ and dimension $M\times 1$; f is a vector composed of discrete particle size distribution with the element of $f(d_i)$ and the dimension $N\times 1$; A_θ is the kernel matrix corresponding to the electric field autocorrelation function data, and the dimension is $M\times N$, and the elements are:

$$A(j,i) = \exp\left(-\frac{16\pi k_B \mathrm{T} n_m^2(\lambda_0)\sin^2(\theta/2)}{3\eta d_i \lambda_0^2}\cdot\tau_j\right) \tag{3.87}$$

The particle size information can be obtained by solving the above equation.

3.3.2 Optical System of Optical Particle Counter

The optical system of an optical particle counter is mainly composed of particle illumination system, scattered light collection system, and transmitted light collection system [27], in which the particle illumination system and scattered light collection system are perpendicular to each other.

3.3.2.1 Particle Lighting System

The role of the particle illumination system is to generate a laser beam of a certain shape, so that particles will be illuminated when passing through the laser beam, generating scattered light. The laser beams are converged, as shown in Figures 3.2–3.7, so that particles pass through the focal point of the lens in turn, and the scattered pulse signal is detected for counting. However, in the actual environment, it is impossible to ensure that all particles pass through the focus position, while the laser at other locations is relatively thick, which is prone to multiple particles being illuminated at the same time, leading to the loss of particle count. In order to improve the counting accuracy, it is a better method to shape the laser from bundle to sheet [28]. Figure 3.8 is the schematic diagram of point laser shaping to plane laser based on plane convex cylindrical mirror. The point laser is compressed after passing through plane convex cylindrical mirror to become a sector laser with a certain divergence angle. To avoid sector laser irradiating the surrounding solid wall, the sector laser can be shaped to parallel laser through a convex mirror, and the particle movement direction is perpendicular to the laser on this surface. In this scheme, the planoconvex cylindrical mirror has the same focal length as the convex lens.

The counting of surface laser generated by plane convex cylindrical mirror meets the counting of particles in most cases. However, since the thickness of surface laser is about 1–2 mm, when the particle number concentration is large, the particles are close to each other, which will cause multiple particles to pass through the surface laser at the same time, as shown in Figure 3.9. The counting of particles is realized by calculating the number of generated pulse signals. When multiple particles are irradiated by laser at the same time, the pulse signals will overlap, resulting in the

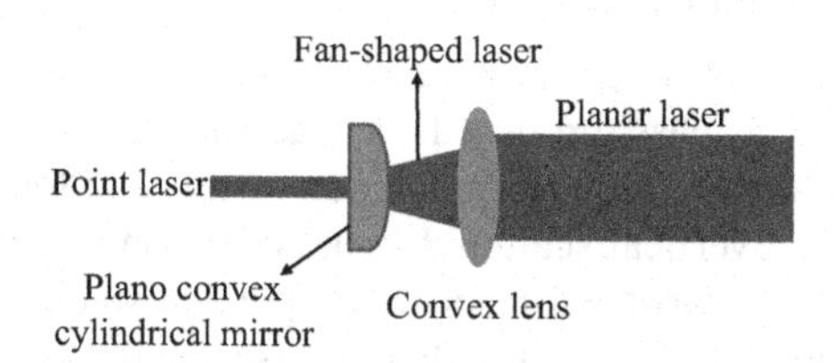

FIGURE 3.8 Point laser shaping to surface laser based on plane convex cylindrical mirror.

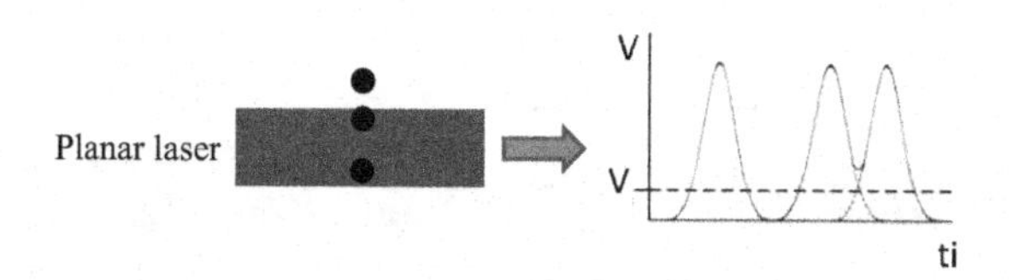

FIGURE 3.9 Particle count loss caused by multiple particles passing through the surface laser at the same time.

particles being counted only once, thus resulting in the loss of particle counting and the decrease of measurement accuracy [29].

In order to avoid particle counting loss and improve counting accuracy, the effective method is to reduce the width of the surface laser so that the laser can only irradiate one particle at the same time, and then the Powell prism shaping laser technology is generated, as shown in Figure 3.10. First, two lenses are used to form a lens group to reduce the spot size of the point laser. The contracted point laser passes through the Powell prism, which can shape the contracted point laser to a thickness of 30 μm. The fan-shaped laser below 30 μm is shaped into a very thin parallel plane laser through a convex lens, and the particle movement direction is perpendicular to the laser plane. Since the laser thickness on this side is below 30 μm, combined with particle dilution technology, it can accurately count 200,000 particles per cubic centimeter.

When particle size inversion is realized based on particle scattering light pulse signal, it is expected that particles with the same particle size will cause the same scattering light pulse signal, which requires uniform spanwise intensity of the surface laser, to avoid the difference of scattering signal intensity caused by different positions of particles passing through the surface laser [30]. Figure 3.11 shows the comparison of the spanwise intensity after the beam shaping of Powell prism and planoconvex cylindrical mirror. The spanwise intensity of the laser after the planoconvex cylindrical mirror shaping is Gaussian distribution. The laser intensity at the center is strong, while that at other locations is weak, which will lead to different pulse signal heights when particles pass through different locations. After the Powell prism is shaped, the laser spanwise intensity is uniform, and the generated scattered light intensity when the particles pass through different positions is consistent, thus avoiding the influence of the particle passing through the surface laser position on the scattered pulse signal intensity.

3.3.2.2 Scattered Light Collection System

Scattered light collection system is used to effectively detect the scattered light generated when particles pass through the surface laser. Currently, the commonly used types are vertical scattered light collection system and forward scattered light collection system. The vertical scattered light collection system is vertical to the transmitted light, easy to arrange in space, simple and stable in system structure, and widely used. The system structure diagram is shown in Figure 3.12. A spherical condenser is arranged in the vertical direction of the transmitted light. The scattered light that shines on its surface is reflected and converged to the other end of the transmitted laser, and a photodiode is arranged at its focus to collect the

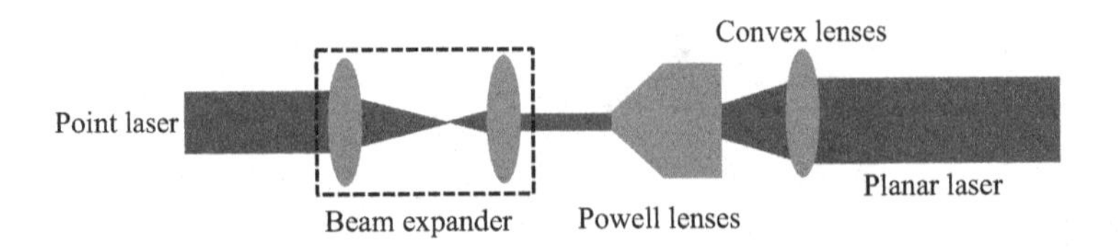

FIGURE 3.10 Point laser shaping to surface laser based on Powell prism.

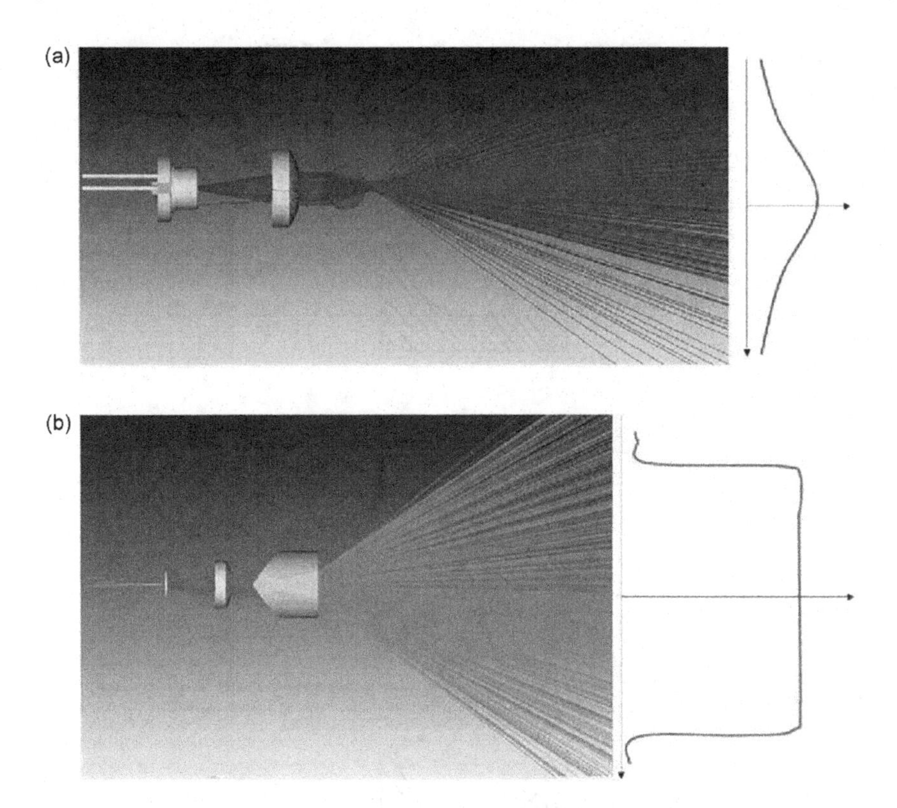

FIGURE 3.11 Comparison of beam shaping between planoconvex cylindrical mirror and Powell prism mirror.

scattered light signal. For some special applications, such as particle concentration detection in high temperature and high-pressure air flow, the photodiode may not work normally. Under such circumstances, lens groups can be used to move back the laser gathered by the spherical condenser, and the first lens will bear and block the high temperature and high-pressure air flow, thus ensuring the normal operation of the photodiode.

Because the forward scattering light intensity of particles is high, and the lateral scattering light intensity of particles with small particle size is weak, it is difficult to detect. In such a case, the forward scattering light may still be effectively detected. Therefore, the forward scattering light collection system has been widely used. The schematic diagram of the forward scattered light collection system is shown in Figure 3.13. Two annular lenses are used to nest the light trap in the center of the annular lens, so as to ensure that the transmitted light is still directly incident into the light trap and collected. The scattered light within the forward angle shines on the first lens and turns into a parallel light. After the parallel light shines on the second lens and converges with the light trap, a photodiode disposed at the focus of the second lens converts the generated scattered light signal collection into an electrical signal.

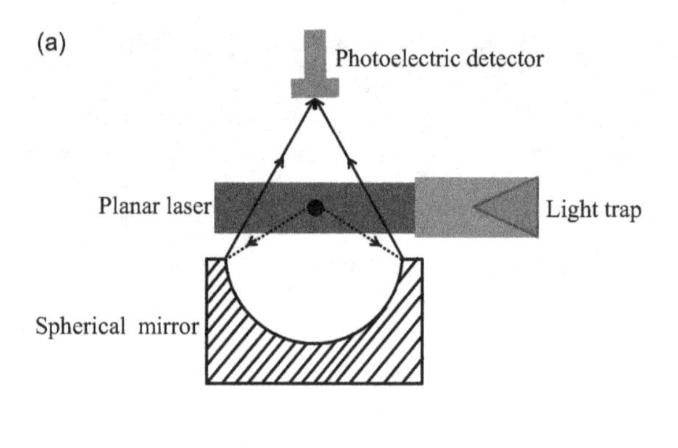

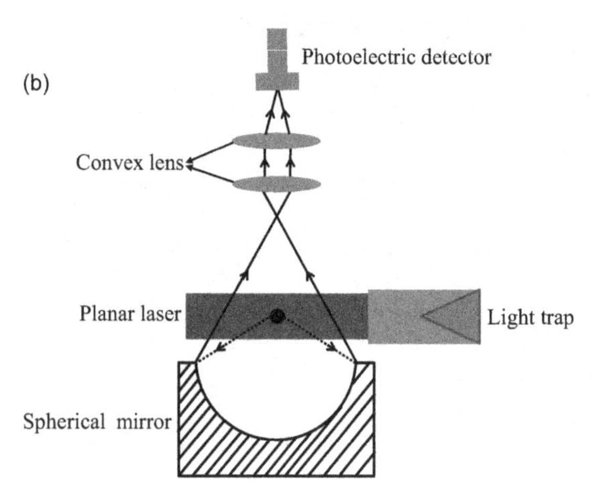

FIGURE 3.12 Vertical scattered light collection system.

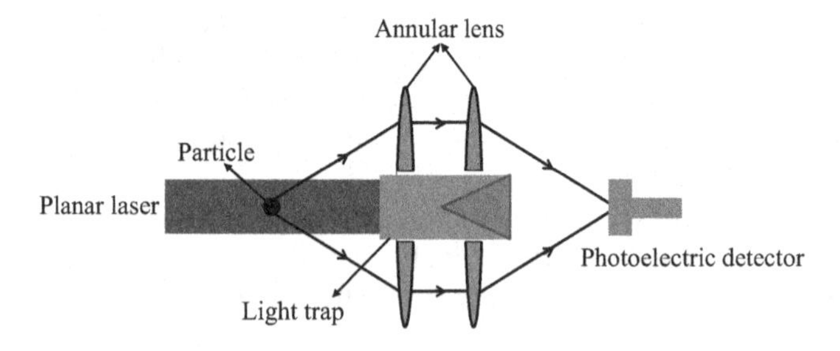

FIGURE 3.13 Forward scattered light collection system.

3.3.2.3 Transmitted Light Collection System

The role of the transmission light collection system is to effectively collect the transmission light, so as to avoid the influence of the transmission light, which is much stronger than the scattering light, on the detection of the scattering signal. The commonly used transmission light collection device is a light trap. Its geometric structure is shown in Figure 3.14. There is a conical structure inside it, which reflects light on the inner wall. The inner wall is equipped with threads to prevent multiple reflections of light, so as to prevent the collected transmission light from "escaping" from the light trap in its internal multiple reflections. The light trap is usually made of aluminum alloy, and its surface is anodized to increase the trapping capacity of redundant light.

3.3.3 Optical Particle Counter Photoelectric Conversion Method

The process of converting the weak scattered light signal in the optical particle counter into an electrical signal is shown in Figure 3.15. The scattered light gathered by the spherical mirror generates a current signal on the photodiode. The current signal is converted into a voltage signal after the transimpedance amplifier. After the voltage signal is amplified, it is sent to the data acquisition equipment for high-speed sampling of the signal to obtain a digital signal, and the digital signal can be processed to obtain pulse data.

Photodiodes are semiconductor devices that convert optical signals into electrical signals. Compared with ordinary diodes, they are structurally different. In order to receive incident light easily, the junction area of photodiodes is larger and the electrode area is designed as small as possible. In addition, the junction depth of photodiodes is very shallow, generally less than 1 μm. Photodiodes work under the action of a reverse voltage. When there is no light, the reverse current is very small (generally less than 0.1 μA), which is called dark current. When there is light, the photons carrying energy enter the PN junction, and transfer the energy to the bound electrons

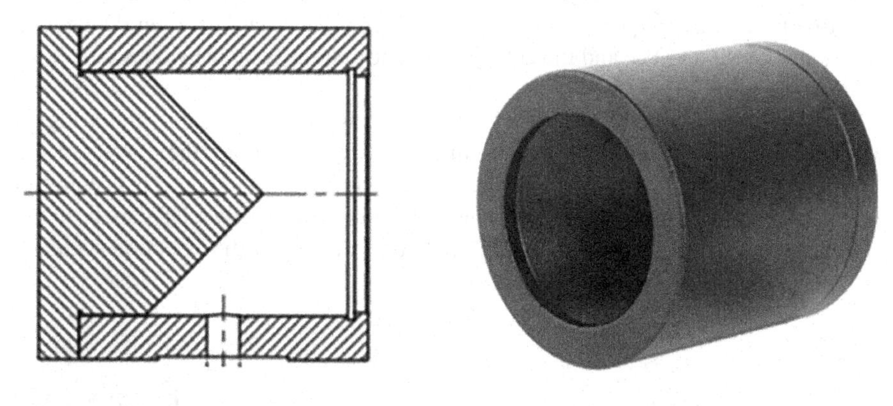

FIGURE 3.14 Optical trap structure.

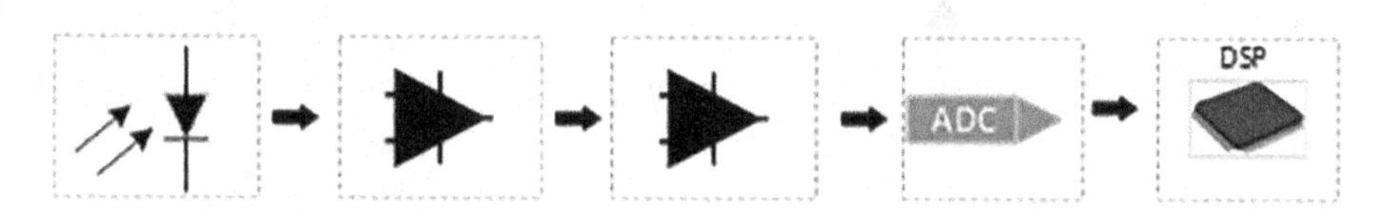

FIGURE 3.15 Optical particle counter optical signal conversion to electrical signal method (From left to right: photodiode, cross-group amplifier, reverse amplifier, analog-to-digital converter, digital signal processor).

on the covalent bond, so that some electrons break free of the covalent bond, thus producing electron-hole pairs, called photogenerated carriers. They participate in the drift movement under the action of reverse voltage, making the reverse current significantly larger. The greater the intensity of light, the greater the reverse current. This property is called "photoconductivity". The current generated by photodiode under the illumination of general illuminance is called photocurrent. If a load is connected to the external circuit, an electrical signal is obtained on the load, and the electrical signal changes with the change of light (Figure 3.16).

Generally, the photodiode should be connected to the power supply reversely, which can increase the photocurrent generated by the photodiode and improve the tube response speed of the photodiode. The selection of photodiode mainly focuses on the central wavelength, operating voltage, photocurrent response speed, and dark current size. It is also necessary to pay attention to the operating temperature of the photodiode, because the photodiode will shift the photocurrent under high-temperature conditions, which causes the shift of signal during sampling.

Transimpedance amplifiers are generally used in high-speed circuits, such as photoelectric transmission communication systems, because of their advantage of high bandwidth. The photocurrent generated by the photodiode is amplified and output through the transimpedance amplifier, which realizes the function of converting the optical signal into an electrical signal and then preliminarily amplifying the electrical signal. The magnification depends on the value of R1 (Figure 3.17).

The signal after transimpedance amplification is a negative signal. In order to facilitate the post stage processing, the negative signal needs to be converted into a positive signal through reverse amplification. Meanwhile, the signal can be further amplified, and low-frequency noise can be filtered. The magnification depends on the values of R2 and R3, and the cut-off frequency depends on the values of R2, R3, and C1 (Figure 3.18).

Analog–digital conversion converts the amplified analog signal into digital signal. Digital analog conversion devices are mainly selected according to the conversion rate and conversion bit. The conversion rate is related to gas flow rate, gas concentration and laser thickness in OPC, and the number of data bits is directly related to the detection accuracy. The higher the number of data bits, the higher the detection accuracy.

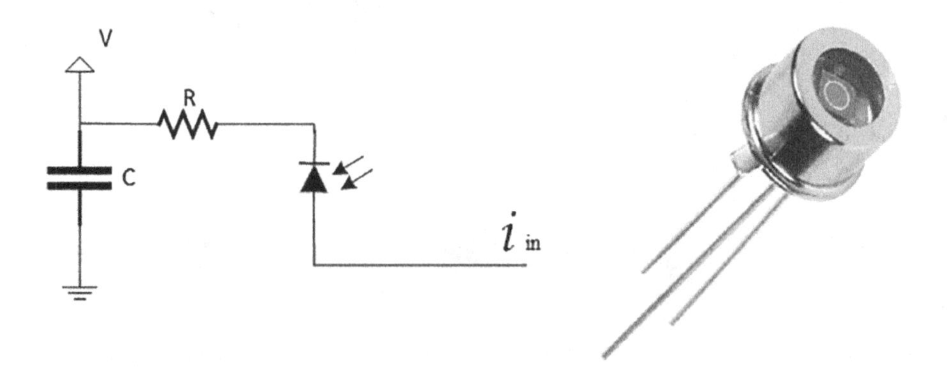

FIGURE 3.16 Circuit principle and physical diagram of photodiode.

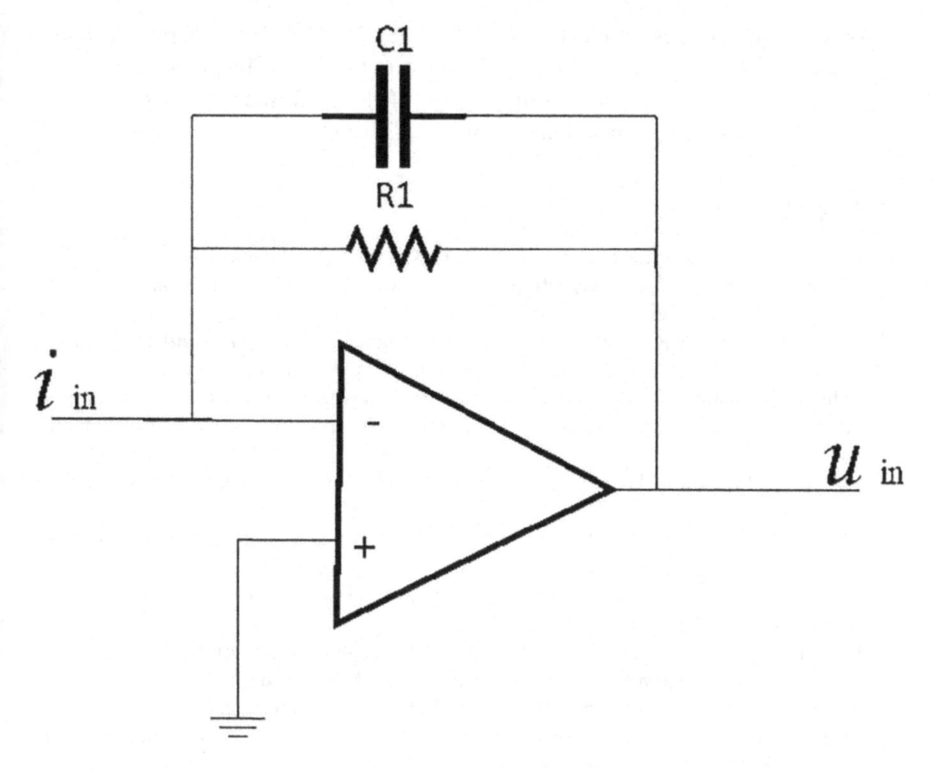

FIGURE 3.17 Schematic diagram of transimpedance amplifier circuit.

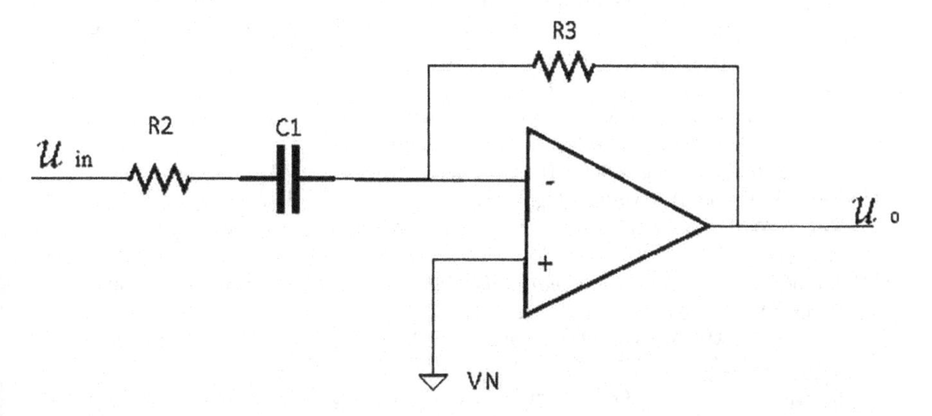

FIGURE 3.18 Schematic diagram of reverse amplifier circuit.

The selection of data bits depends on the laser light intensity, particle size, and scattered light intensity. Digital signal processor is mainly used to process and analyze the collected digital signals, such as digital filtering, particle counting, statistical analysis, etc.

As one of the most essential apparatuses in the CPC (condensation particle counter) system, OPC will serve as the optical counting function. The precision and efficiency of OPC will be the determining factor of the performance of CPC system, which will be further introduced in the following chapter.

REFERENCES

[1] Fomin V M, Pokatilov E P, Devreese J T, Klimin S N, Balaba S N, Gladilin V N. Proceeding of the 23rd International Conference *on the Physics of Semiconductors.* Singapore: World Scientific, 1996: 1461.

[2] Baron PA, Willeke K. *Aerosol Measurement: Principles, Techniques, and Applications.* 2nd ed. United States, 2001. New Jersey: John Wiley & Sons, Inc.

[3] Haines M, Scamarcio G. *Phonons in Semiconductor Nanostructures*, Vol. 236 of *NATO Advanced Studies Institute, Series E: Applied Sciences.* Holland: Kluwer Academic, 1993: 93.

[4] Hayes W, Loudon R. *Scattering of Light by Crystals.* New York: John Wiley and Sons Inc., 1978

[5] Hellwege K H. *Physics of II-V and I-VII Compounds, Semimagnetic Semiconductors.* Vol. 17 of *Landolt-Bormstein New Series.* Berlin: Springer, 1982.

[6] Kartheuser E. *Polarons in lonic Crystals and Polar Semiconductors.* Ed by Devreese J T. Amsterdam: North-Hllnd, 1972: 728.

[7] Mitchell D L, Palik E D, Zemel J N. Proceeding of the *7th* International Conference *on Physics of Semiconductors.* Ed by Hulin M. Paris: Dunod, 1964: 325.

[8] Wolf E. *Principles of Optics.* 6th ed. New York: Cambridge University Press, 1997.

[9] Carin L. *Ultra-Wideband, Sbort-Pulse Electromagnetics.* Ed by Bertoni H. New York: Plenum Press, 1993.

[10] Stanley H E. *Introduction to Phase Transitions and Critical Phenomena.* New York: Oxford University Press, 1971: 98.

[11] Sheng J, Zhang D. Mie scattering study on the formation of phase structure and phase size distribution of concatenate alloys. *Journal of Light Scattering*, 2003, 14: 252.

[12] Bohren C F, Huffman D R. *Absorption and Scattering of Light by Small Particles.* New York: John Wiley and Sons, 1983.

[13] Mishchenko M I, Travis L D., Lacis A A. *Scattering. Absorption and Emission of Light by Small Particles.* Cambridge: Cambridge University Press, 2002.

[14] Van de hulst H C. Lighr Scattering by Small Particles. New York: Dover, 1981.

[15] Berne B, Pecora R. *Dynamic Light Scattering.* New York: John Wiley and Sons, 1976.

[16] Kerker M. *The Scattering of Lighrand Other Electromagnetic Radiation.* New York: Academic, 1969.

[17] Minneart M G J. *Light and Color in the Outdoors.* New York: Springer-Verlag, 1993.

[18] Pesic P. *Sky in a Bottle.* Cambridge, MA: MIT Press, 2005.

[19] Sorensen C M, Fischbach D E. Patterns in Mie scattering. *Optics Communications*, 2000, 173: 145–153.

[20] Sorensen, C M, Shi D. Guinier analysis for homogeneous dielectric spheres of arbitrary size. *Optics Communications*, 2000, 178: 31–36.

[21] Saarnio K, Teinila K, Aurela M. High-performance anion-exchange chromatography-mass spectrometry method for determination of levoglucosan, mannosan, and galactosan in atmospheric fine particulate matter. *Analytical and Bioanalytical Chemistry*, 2010, 398(5): 2253–2264.

[22] Su M, Xu F, Cai X, Ren K, Shen J. Optimization of regularization parameter of inversion in particle sizing using light extinction method. *China Particuology,* 2007, 5(4): 295–299.

[23] Ma L, Hanson R K. Measurement of aerosol size distribution functions by wavelength-multiplexed laser extinction. *Applied Physics B: Lasers and Optics*. 2005, 81(4): 567–576.

[24] Hulst V. *Light Scattering by Small Particles*. London: Chapman and Hall, 1957.

[25] Luo J, Trampe A, Fissan H. A new particle counter using non-imaging optics and real-time correlation filter for particle detection. *Aerosol Science and Technology*. 1999, 30: 545–555.

[26] Schatzel K. Correlation techniques in dynamic light scattering. *Applied Physics B*, 1987, 42(4): 193–213.

[27] Hove D, Self S. Optical sizeing for in-situ measurements. *Applied Optics*, 1979, 18(10): 1632–1645.

[28] Nourpour N, Olfert J. Calibration of optical particle counters with an aerodynamic aerosol classifier. *Journal of Aerosol Science*, 2019, 138: 105452.

[29] Wu T, Murashima Y, Sakurai H, Iida K. A bilateral comparison of particle number concentration standards via calibration of an optical particle counter for number concentration up to ~1000 cm− 3. *Measurement*, 2022, 189: 110446.

[30] Shin D, Woo C, Hong K, Kim H. Continuous measurement of PM10 and PM2.5 concentration in coal-fired power plant stacks using a newly developed diluter and optical particle counter. *Fuel*, 2020, 269: 117445.

4 Condensation Particle Counter

Zheng Xu, Kang Pan, Lei Liu, Shanshan Tang, and Zichen Zhang

4.1 INTRODUCTION

In the current market, there are several types of condensation particle counters (CPCs) that include the laminar-flow type, mixing type, and expansion type [1,2]. The fundamental principles and the particle detection systems of these CPCs do not have significant differences, but the methods to achieve fluid supersaturation are distinctive. A continuous laminar-flow CPC, as shown in Figure 4.1, is the most common type of CPCs due to its (relative) technological simplicity. During operation, it allows a supersaturated vapor to condense on the aerosol particles and then makes them grow to sizes that can be optically detected [3–7]. In a laminar-flow CPC, the saturator, condenser, and optical particle counter (OPC) are the three key components. The saturator is used to heat the working fluid to achieve supersaturation, which will be filtered to remove any contaminants inside before entering the condenser. In the condenser, the airflow is quickly cooled down to a preset temperature, which enforces liquid condensation on the particle surface. This leads to particle

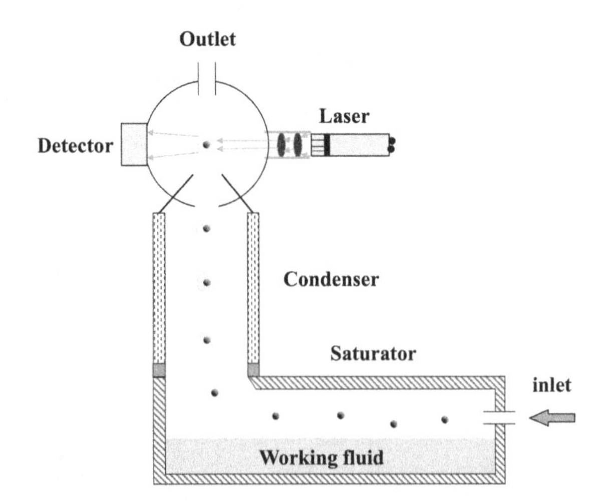

FIGURE 4.1 Schematic diagram of the laminar-flow CPC.

DOI: 10.1201/9781003423195-4

growth from several nanometers to several micrometers, which is the minimum size that can be detected by the OPC. The condensed particles are then injected one-by-one from a nozzle into the OPC, where they are counted using light scattering techniques [3–7]. This chapter will provide a detailed introduction to the three core components, namely the saturator, condenser, and OPC, and other auxiliary systems, such as air transportation system, air filter, and air pump, within a CPC.

As the key component of a CPC, the saturator is specially designed to generate a supersaturated gas in which fine particles are suspended by heating the working fluid. Typically, a saturator comprises a heating pipe, working fluid, and a reservoir. The reservoir is used to contain the working fluid, which is heated by the heating pipe to enforce its evaporation and then supplied to the condenser [4,8–10]. Saturator is usually made of aluminum alloy in the shape of a circle or a square. The heating wires (e.g., Nickel-chromium heating resistance wire) are mounted on the heating pipe, which is usually positioned under the reservoir to ensure uniform heating, by monitoring the temperature of the heating pipe using thermocouples. In the saturator, n-butanol is usually chosen as the working fluid immersed in a porous media, which are usually silicon carbide and aluminum foam. This selection not only increases the interface area between air and working fluid but also provides an ample supply of saturated working fluid to the condenser [4,7].

Condenser is another important component of a CPC, which usually consists of a cooling system, a contraction section, and a nozzle. Once the particles enter the condenser, condensation occurs on the particle surface to enlarge its size to an optically detectable value [4,8–10]. At the end of the condenser, the condensed particles are then injected into the OPC. Both the inner diameter and length of the condenser can significantly affect the condensation process. The inner diameter affects mass and heat transfer within the working fluid between the condenser wall and the center, where most of the particles are located. A larger inner diameter increases the time required to achieve supersaturation within the condenser. The length of the condenser determines the residence time of suspended particles, with shorter condensers leading to smaller diameters of condensed particles due to their rapid passage through. The temperature control is another important factor that affect the condensation rate. The condenser's temperature is reduced by cooling fins, which are often connected to the wall of CPCs. The heat generated by cooling fins is dissipated by powerful fans. In addition, the nozzle, mounted at the end of the condenser, controls the duration for which particles remain in the particle counter. The maximum particle concentration detected by the particle counter is greatly influenced by this duration. Adjusting the nozzle diameter can regulate the injection velocity of particles, with a smaller nozzle increasing the injection velocity and allowing for a higher aerosol concentration. Furthermore, the distance between the nozzle and the OPC should be specially designed, as it determines the particle motion after injection. If the distance is too large, the particles can stay around the OPC instead of passing through quickly and directly.

Another important part of a CPC is the OPC, consisting of the laser, optics, and photodetectors. In an OPC, the condensed particles scatter the emitted laser when passing through, and the scattered light is then focused by the optics to a photodetector and then converted to the electrical signal to count the particle number [4,8–10]. For a 0.5 mm particle, the electrical signal generated by the photodetector in the OPC is only around 100 mv, and hence, it is necessary to amplify the signal conditions first before comparing with the reference voltage, which can be set by the microcontroller with a typical value between 2.5V and 5V. The lower limit of the OPC is able to be changed by adjusting the magnification through the pulse signal. As the size of the condensed particles can reach microns, it is only necessary to ensure a lower limit of less than 1 μm, during the measurement using the OPC.

For a conventional CPC, it also includes some other important components, in addition to the three core parts. The first one is a sampler, whose conventional structure is shown in Figure 4.2, that the gas sample flows vertically into the prototype system through the intake pipe. In the sampler, a very short part of the sampling capillary penetrates vertically into the intake pipe, through which a small part of the sample gas enters the inlet of the condenser. In the meanwhile, most of the remaining gas is discharged as exhaust through the exhaust pipe of the sampler. The structural design scheme of the sampler needs to ensure the isokinetic sampling of the sample gas, meaning that particle number concentration within the gas in the capillary tube should be the same as those inside the sample gas.

In addition, the working mechanism of CPCs determines only a small range of 20–40 mL/min sample gas flowing into the condenser. In a CPC, the condensed particles should pass through the OPC one by one, which requires both a low particle concentration and a small flow area to ensure only one particle in the flow volume with a small distance. As shown in Figure 4.2, the sampler designed can ensure a

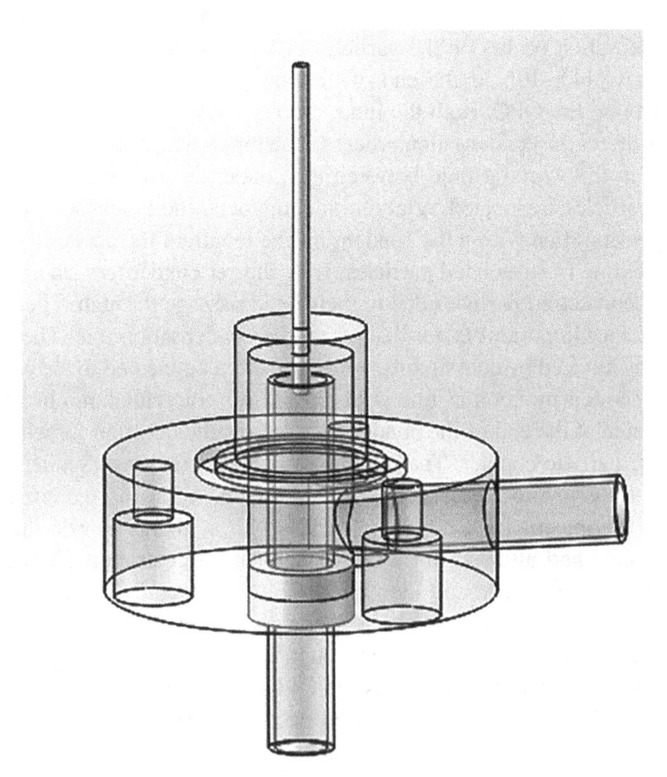

FIGURE 4.2 Schematic of a sampler in CPCs [9].

large amount of sample passing through the inlet of the OPC. Compared to a small gas flow, the large flow volume by this sampling method is able to increase the response time and reduce the loss of particles [11], and it also avoids a complex gas path and reduces the error during the flow rate measurement.

CPCs also include other components, such as air filter, air pumps, flow meter, and valves. Air filters are used to remove contaminants in the air to avoid the error of counting particles and also to reduce the damage to the air pump. Air pumps can provide enough power for the air path to transport particles, and they can work together with other auxiliary equipment (such as flow meter and valves) and the electronic signal control system, to control and adjust the important parameters (such as gas flow) in the counting system. As a result, these auxiliaries will affect the particle counting efficiency and accuracy of a CPC as well [8–10].

4.2 SATURATOR SECTION

4.2.1 Structure of Saturator

The saturator section is the main part of the particle nucleation unit within the condensation particle counting system. Its primary function is to saturate a certain portion of the airflow with vapor from the working fluid [12]. This vapor is then used to grow particles into optically detectable droplets in the subsequent condenser. Typically, a saturator consists of several key parts, including the working fluid, fluid reservoir, fluid-soaked porous medium, and heater. In the working process, the fluid reservoir functions as a storage for the working fluid. To facilitate the generation of fluid vapor, the saturator is heated by a heater to preset temperatures that correspond to the selected working fluids. Precision thermistors are utilized for temperature measurement within the saturator. The working fluid vapor is then transported to the condenser by carrier gas. Typically, the saturator is constructed using metallic materials, such as aluminum alloy, and its structure can be designed in round or square shapes based on specific requirements. The heater is generally composed of multiple heating rods (e.g., nickel-chromium heating resistance wire), which are evenly arranged around the saturator to ensure uniform heating. In the saturator section, a fluid-soaked porous medium is commonly employed. This medium serves to increase the contact area between the airflow and the working fluid, which can effectively enhance the efficiency of the working fluid being carried by the airflow. This ensures that the gas at the outlet of the saturator remains saturated even when the fluid level is low. This arrangement ensures a sufficient supply of saturated working fluid vapor for the condenser.

4.2.2 Working Fluid

The working fluid is used to generate saturated vapor at a higher temperature in the saturator and condense on the surface of ultrafine particles at a lower temperature in the condenser to realize the condensation growth of particles [12]. The choice of working fluids significantly impacts the process of condensation growth. Several key factors, including wettability, surface tension, diffusion coefficient, and thermal stability, are essential in determining the appropriate working fluid selection [13]. The contact angle is used to characterize the wettability of working fluid with particles.

Different working fluids exhibit varying contact angles with particles, thus influencing the activation efficiency of particles based on the heterogeneous nucleation theory. The smaller the contact angle between the working fluid and particles, the easier it is for the working fluid to condense on the surface of the particles [14]. As a result, a lower level of supersaturation in the working fluid vapor is required for nucleation, reducing the probability of homogeneous nucleation. Additionally, the surface tension of working fluids affects the Kelvin diameter of the activated particles. A lower surface tension of the working fluid results in a smaller minimum Kelvin diameter required to maintain the stability of activated particles [15]. Therefore, the selection of working fluid with weaker surface tension can reduce the minimum size of activated particles and improve the counting efficiency of the CPC.

According to the properties of condensation working fluids, there are two main types commonly used: alcohol and water [4]. Among the alcohols, n-butanol, isopropanol, and diethylene glycol are frequently utilized. Comparatively, isopropanol and diethylene glycol exhibit better condensation growth capacity than n-butanol. This is because n-butanol tends to undergo homogeneous nucleation at high supersaturation levels. However, isopropanol and diethylene glycol have higher water solubility than n-butanol. In measurements of atmospheric aerosols with high humidity, water is more likely to diffuse and dissolve into the solutions of isopropanol or diethylene glycol. This has an impact on the condensation growth of particles and may result in the accumulation of water within the device during long-term operation. Consequently, isopropanol and diethylene glycol are not preferred choices in such scenarios.

n-Butanol is widely used as the primary working fluid in CPCs for several reasons. Firstly, n-butanol is a macromolecular substance with a low mass diffusion coefficient. This characteristic allows it to remain in the vapor stream, leading to a stable and repeatable condensation phenomenon. Moreover, it has a lower tendency for water to diffuse and dissolve into it, making it more suitable for long-term usage. In an n-butanol-based CPC, a warm, wet-wall saturator and a chilled-wall condenser are commonly employed (Figure 4.3) [16,17]. This design is due to the significant disparity between the thermal diffusivity of air (0.215 cm^2/s) and the vapor diffusivity of n-butanol (0.081 cm^2/s). As the n-butanol vapor follows the airflow into the condenser, the n-butanol vapor diffusion rate is much smaller compared to the air temperature reduction rate, and therefore, the thermal flux of the gas to the chilled wall occurs at a much greater rate than the butanol vapor flux to the wall. This causes the partial pressure of n-butanol vapor to significantly exceed the equilibrium saturated vapor pressure. Consequently, a supersaturated condensation occurs within the central core of the flow.

The water-based CPC has been gradually developed in the last decade [18–20]. In certain cases where the measurement of specific particles or operating conditions prohibits the use of flammable and explosive substances, the alcohol-based method cannot meet the measurement requirements. Water, on the other hand, proves to be a more suitable working fluid in such applications due to its environmental friendliness, nonflammability, and overall safety [21]. The diffusivity of water vapor in air (0.265 cm^2/s) is approximately 1.3 times of the thermal diffusivity of air (0.215 cm^2/s). As a result, the flux of water vapor to the walls occurs at a faster rate compared to the thermal flux of the gas. This leads to the formation of supersaturation regions

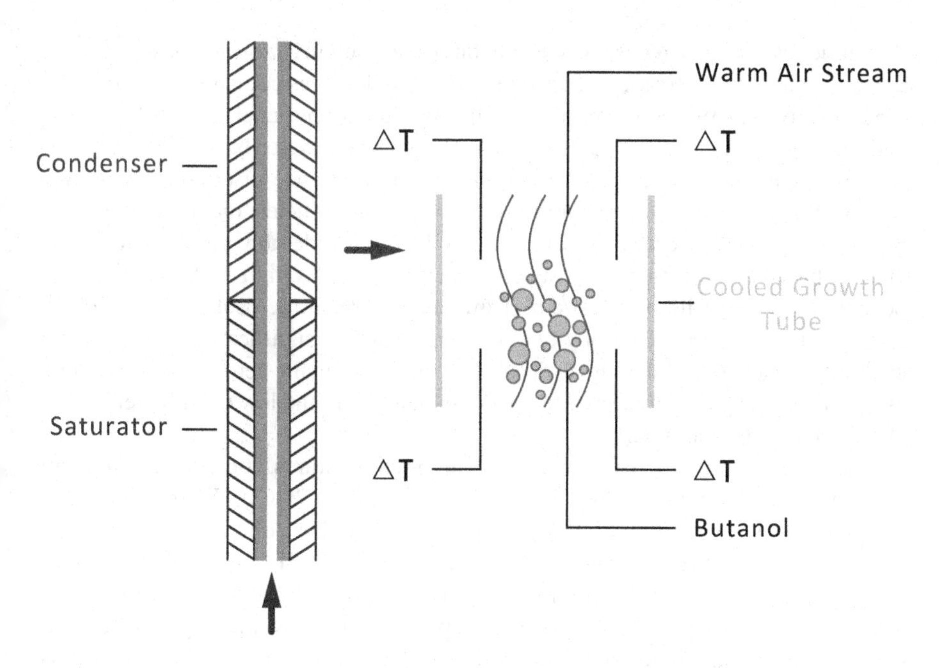

FIGURE 4.3 Schematic diagram of the n-butanol-based condensation particle growth.

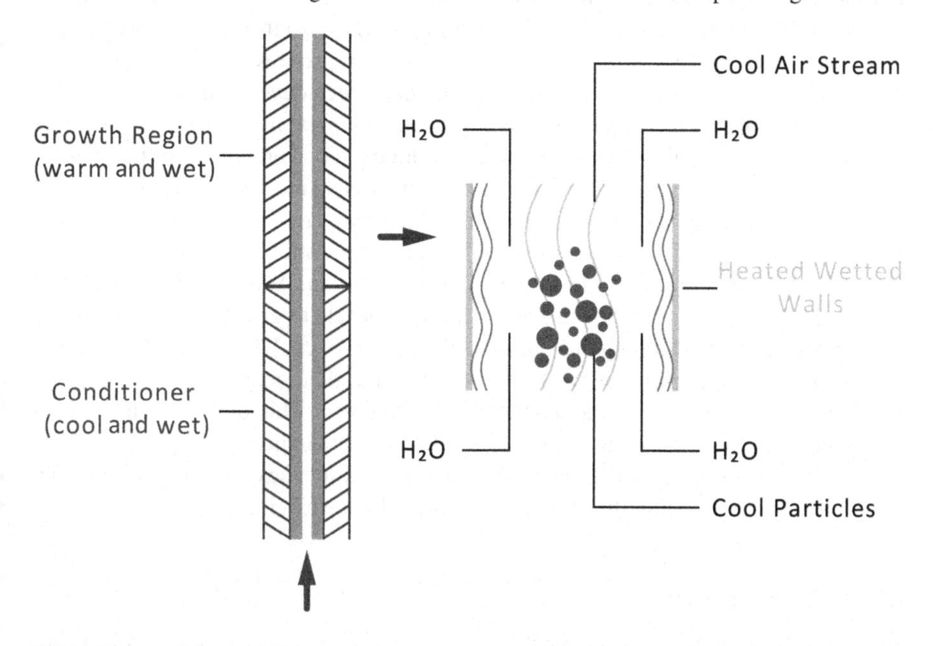

FIGURE 4.4 Schematic diagram of the water-based condensation particle growth.

near the wall surface, while the majority of the flow passing through regions of lower supersaturation results in a reduced particle growth rate and incomplete activation [21]. A clever design, depicted in Figure 4.4, in which a cold saturator is followed by a warm condenser, was proposed, and the first water-based CPC was verified [22].

Theoretically, using water as the working fluid can result in a larger peak supersaturation and enable the activation of particles with smaller sizes compared to commonly used alcohols [14,19]. However, due to the significant physicochemical differences and varying particle affinity between alcohols and water, water-based CPC and alcohol-based CPC can sometimes exhibit notable performance discrepancies. Biswas et al. [20] and Hering et al. [22] reported that for salt particles, the number concentration of particles detected by water-based CPC is considerably higher than that by butanol-based CPC, with a counting efficiency reaching as high as 97% at the number concentration of 10^5 particles/cm^3. However, for organic aerosol particles and oily particles, the counting efficiency of water-based CPC is significantly lower. Romay et al. [21] demonstrated that the hydrophobic surface of many aerosol particles in the air makes it impossible for water to condense, leading to the low counting efficiency observed in water-based CPC.

The exhaust emitted from engines at high-temperatures contains a significant amount of water vapor and volatile organic compounds (VOCs). When this exhaust is cooled in the exhaust pipe, these compounds can either adhere to the surface of solid particles or directly condense into liquid particles. If a standard room temperature CPC is used for measurement, this can result in significant counting errors. Therefore, accurately measuring solid particles emitted by the engines becomes challenging. In accordance with EU regulations for measuring particulate emissions, the standard measurement method requires maintaining the temperature of sampled gas above 150°C before particles sampling to prevent condensation of volatile substances. To address this issue of condensation of VOCs and water vapor during the measurement process, high-temperature CPCs have been developed.

The commonly used working fluids in room temperatures, such as water and alcohols, have low boiling points and high saturated vapor pressures. Consequently, they cannot be utilized in high-temperature conditions since they are incapable of infiltrating and encapsulating particles at elevated temperatures, thereby preventing nucleation. The working fluids employed should demonstrate physicochemical stability at high temperatures and in the presence of oxygen. Additionally, they should possess higher boiling points and lower saturation levels to ensure slow evaporation during operation, minimizing fluid waste. Addressing the challenge of direct counting of nanoparticles at high temperatures, a new working fluid consisting of high Lewis number oil and near-azeotropic alcohols was developed. This development enabled direct condensation counting of nanoparticles at the temperature of 300°C [23]. Moreover, silicone oils, perfluorocarbons, and perfluorotrihexylamine are potential high-temperature working fluids that required further research in the future.

4.2.3 Porous Media in CPC

Porous media refers to a solid material that contains numerous tiny holes. These holes have a certain degree of connectivity, allowing the fluid to flow through them under specific conditions. There are various types of pore structures, and their interconnected relationships are quite intricate. The pore structure plays a decisive role in the structural characteristics, mechanical properties, and fluid flow in the porous

FIGURE 4.5 Pictures of porous medium. (a) silicon carbide; (b) aluminum foam; (c) porous Teflon.

medium. Consequently, it also affects the practical utility of the porous medium. It is vital for the porous medium to exhibit good wettability toward the working fluids. This enhances the contact area between the working fluid and the carrier gas, consequently increasing the overall evaporation rate of the working fluids within the saturator [12]. Such enhanced wettability ensures an ample supply of adequately saturated working fluid steam. The commonly used porous media in CPC saturator includes silicon carbide (Figure 4.5a), aluminum foam (Figure 4.5b), and Teflon (Figure 4.5c).

4.2.4 Numerical Simulation of Designed CPC Saturator

In order to validate the feasibility of CPC design, numerical simulation using computational fluid dynamics (CFD) are commonly employed. CFD is an interdisciplinary field that combines fluid mechanics and computer science. It involves approximating the integral and differential terms in the governing equations of fluid mechanics, converting them into discrete forms, and solving these algebraic equations using computers to obtain numerical solutions in time or space. CFD has emerged as a widely utilized tool for structural design and performance simulation of CPC systems. By accurately modeling and calculating the particle growth process within CPC, it enables precise prediction and analysis of particle growth under time-saving and labor-efficient conditions. Compared to experimental measurements, obtaining flow field data in CPC is easier through CFD. The advancements in CFD have significantly facilitated theoretical research and numerical simulation of CPC systems.

Wang et al. [17] developed a two-dimensional axisymmetric model in the COMSOL Multiphysics simulation software to investigate the spatial distribution of the saturation ratio in the designed CPC. The model was also used to predict the minimum diameter of particles that can be activated by condensation of n-butanol vapor. They also studied the non-uniform spatial distribution of internal temperature, vapor partial pressure, and saturated vapor pressure under specific conditions, including a saturation chamber temperature of 35°C, condensation chamber temperature of 10°C, inlet aerosol temperature of 35°C, condensing chamber diameter of 4.5 mm, condensing chamber length of 9.7 cm and sampling flow rate of 0.3 L/min [16,17]. It is proved that the temperature distribution within the saturator is uniform.

4.3 CONDENSER

4.3.1 Structure and Principle of the Condenser Segment

The condenser segment stands as the pivotal structure within the nuclear particle counter for achieving the growth of particles through condensation [19–21]. A visual representation of the physical model can be observed in Figure 4.6. The critical design dimensions of the condenser segment may be ascertained through a mathematical model, with the diameter and length of the internal channel emerging as the most consequential structural parameters influencing the efficiency of particle counting [14]. The diameter of the internal channel in the condenser segment exerts an impact on the heat and mass transfer of the working medium vapor along the wall and the central axis, where the particle number concentration is most significant. As the diameter increases, the working medium vapor requires more time to attain the state of supersaturation at the central axis [22]. On the other hand, the length of the condenser segment affects the retention time of particles [17]. A shorter length implies reduced time, leading to a smaller final diameter of particles. In the presence of a specific diameter and length within the internal channel, particles can effectively condense and undergo growth [24–26].

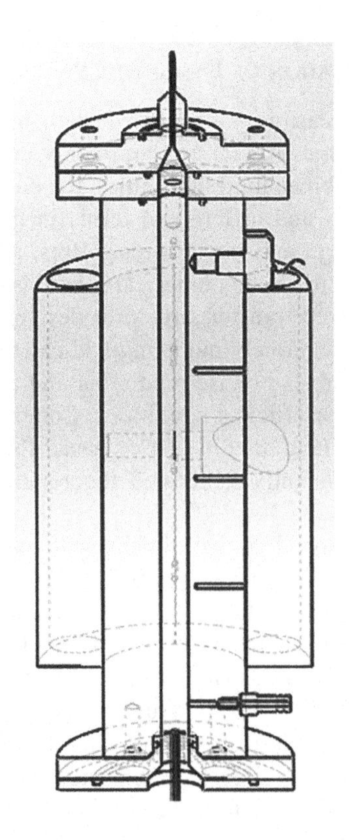

FIGURE 4.6 Physical model of condenser segment.

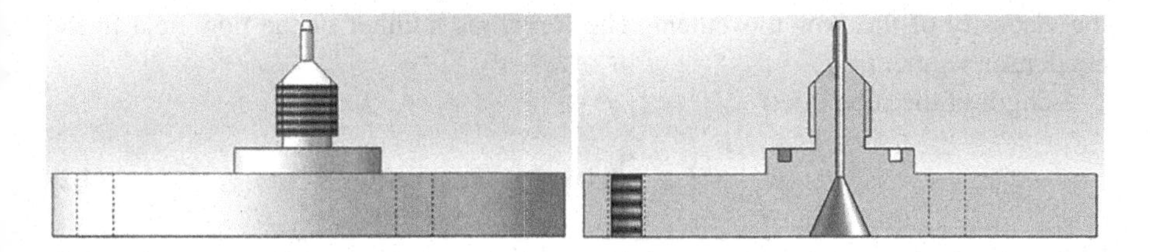

FIGURE 4.7 Nozzle structure.

The condenser segment can be crafted from a robust aluminum alloy material [27]. Due to the extended nature of the inner hole, it takes the form of a longitudinal bore. Consequently, when subjected to conventional drilling processes at both ends, strict coaxiality demands are essential. Additionally, the surface roughness of the inner wall of the condenser segment holds sway over the laminar flow and convective heat and mass transfer, thus necessitating elevated requirements for surface smoothness.

To ensure that the grown particles enter the OPC individually and sequentially, it is imperative to maintain an exceedingly small cross-sectional area of airflow [28,29]. To accomplish this, a nozzle structure, as depicted in Figure 4.7, has been devised. At the nozzle's outlet, a tapering configuration is implemented to reduce the cross-sectional area of the airflow substantially, thereby achieving a diminutive value. Moreover, in order to prevent the airflow from dispersing outward upon entering the photosensitive area of the OPC, the distance between the nozzle outlet and the photosensitive area can be thoughtfully designed.

4.3.2 Mathematical Model of Condenser Segment

The airflow within the condenser segment traverses the central cylindrical channel, with several heating rods symmetrically distributed around its circumference. To maintain a constant temperature of the inner wall of the condenser segment, a temperature sensor is positioned on its upper part [30]. The end of the channel inside the condenser segment serves as the inlet for the gas to be measured, along with the saturated working medium vapor. The capillary tube located at the center of the bottom functions as the inlet for the sampling gas, while the sheath gas (saturated working medium vapor) enters from the surrounding area of the capillary tube. The volume flow of the sheath gas is carefully regulated to enter the saturated segment, while the volume flow of the sample gas enters the capillary.

Assuming that the physical parameters of the gas in the condenser segment remain constant and that viscous dissipation is negligible, the Reynolds number (Re) governing the flow within the condenser segment is as follows:

$$Re = \frac{UD_C}{\nu} \tag{4.1}$$

Whereas D_C denotes the inner diameter of the channel in the condenser segment, U represents the average flow velocity within the condenser segment, and ν signifies the viscosity of the flow movement. The Reynolds number of the flow field in the condenser segment.

Length of the tube inlet:

$$L_{ent} \approx 0.06 \cdot Re \cdot D_c \tag{4.2}$$

In this equation, considering the diameter of the condenser segment and the length of the inlet, we can approximate the length of the condenser where the flow can be regarded as fully developed laminar flow. In this state, there is no mixing of sampling gas and sheath gas, and it does not affect the heat and mass transfer of the saturated working medium steam. Due to the cylindrical flow having uniform circumferential parameters in the condenser segment, we can simplify the calculation of the flow as that of a two-dimensional pipeline. Consequently, the velocity field of the fully developed laminar flow in the condenser segment is as follows:

$$u_z = 2U\left[1-\left(\frac{r}{R}\right)^2\right] \tag{4.3}$$

Where U represents the average airflow velocity in the condenser segment, and R denotes the channel radius in the condenser segment.

The energy conservation equation in the condenser segment is:

$$\rho C_p\left(\frac{\partial T}{\partial t}+u_r\frac{\partial T}{\partial r}+u_\varphi\frac{\partial T}{\partial \varphi}+u_z\frac{\partial T}{\partial z}\right)=\lambda\left[\frac{1}{r}\frac{\partial}{\partial r}\left(r\frac{\partial T}{\partial r}\right)+\frac{1}{r^2}\frac{\partial^2 T}{\partial \varphi^2}+\frac{\partial^2 T}{\partial z^2}\right]+\mu\Phi_V \tag{4.4}$$

Since the system is in a steady state and the circumferential parameters remain constant, we can infer that $\frac{\partial T}{\partial t}=0$, $u_\varphi\frac{\partial T}{\partial \varphi}=0$, $\frac{1}{r^2}\frac{d^2T}{d\varphi^2}=0$; Additionally, since the radial velocity is zero, it follows that $u_r\frac{\partial T}{\partial r}=0$ Moreover, due to the absence of any internal heat source within the condenser segment, we have $\mu\Phi_V=0$. Consequently, the equation is simplified as follows:

$$u_z\frac{\partial T}{\partial z}=\alpha\left[\frac{1}{r}\frac{\partial}{\partial r}\left(r\frac{\partial T}{\partial r}\right)+\frac{\partial^2 T}{\partial z^2}\right] \tag{4.5}$$

Where α is the thermal diffusivity.

Take $\theta=\frac{T-T_w}{T_{0-T_w}}$, $Y=\frac{r}{R}$, $Z=\frac{z}{R\times Pe}$, and make the equation (4.5) dimensionless:

$$\left(1-Y^2\right)\frac{\partial\theta}{\partial Z}=\frac{1}{Y}\frac{\partial}{\partial Y}\left(Y\frac{\partial\theta}{\partial Y}\right)+\frac{1}{Pe^2}\frac{\partial^2 T}{\partial Z^2} \tag{4.6}$$

Where Pe represents the Peclet number, given by $Pe = \frac{D_C U}{\alpha}$, where α is a thermal diffusivity constant. The Peclet number is defined as the ratio of the energy transport rate generated by convection to the energy transport rate generated by heat conduction. In the condenser segment, convection is the primary mode of heat transfer, and thus $Pe >> 1$. As a result, the thermal conductivity term in the equation can be neglected. Consequently, the final temperature governing equation for the flow field in the condenser segment is as follows:

$$u_z \frac{\partial T}{\partial z} = \alpha \left[\frac{1}{r} \frac{\partial}{\partial r} \left(r \frac{\partial T}{\partial r} \right) \right] \tag{4.7}$$

Likewise, by applying Fick's diffusion law and mass conservation law, we can derive the governing equations for the vapor pressure distribution and particle concentration distribution of the working medium within the condenser segment

$$u_z \frac{\partial p}{\partial z} = D_v \left[\frac{1}{r} \frac{\partial}{\partial r} \left(r \frac{\partial p}{\partial r} \right) \right] \tag{4.8}$$

$$u_z \frac{\partial C}{\partial z} = D_p \left[\frac{1}{r} \frac{\partial}{\partial r} \left(r \frac{\partial C}{\partial r} \right) \right] \tag{4.9}$$

In these equations, p denotes the partial pressure of the working medium vapor, C stands for the particle concentration, D_v i signifies the mass diffusion coefficient of the working medium vapor molecules into the gas, and D_p represents the diffusion coefficient of the particles. The Peclet number of the working medium vapor and the particle concentration are denoted as Pe_v and Pe_p, respectively:

$$Pe_v = \frac{D_c U}{D_v} \tag{4.10}$$

$$Pe_p = \frac{D_c U}{D_p} \tag{4.11}$$

Where $Pe_v, Pe_p \gg 1$.

The flow field within the condenser segment exhibits symmetry along the central axis, leading to a temperature gradient of $\partial T/\partial r = 0$ at the axis of the condenser segment (i.e., $r = 0$). Multiple heating rods uniformly heat the condenser segment in the circumferential direction, rendering the heat loss negligible at the upper and lower inlet and outlet connections of the condenser segment. Consequently, the inner wall temperature of the condenser segment can be treated as uniform and constant.

Furthermore, due to the use of a heat-insulating material with low thermal conductivity between the connecting segment of the condenser and the saturated segment, the temperature of the connecting segment closely approximates the temperature of the saturated segment. Additionally, the time taken for the sheath gas to pass through the connecting segment is very short, allowing us to ignore partial pressure loss and heat loss of the steam in the connecting segment.

For simplicity, it is considered that the airflow temperature and saturated steam pressure at the inlet of the condenser segment are equivalent to those of the working medium steam in the saturated segment. A thin layer of working medium liquid film will condense on the condenser wall, leading to the partial pressure of saturated vapor at the wall, matching the partial pressure of saturated vapor at the current temperature.

Regarding the particle flow, it is assumed that the number concentration of particles in all regions except the outlet of the capillary tube is zero. At the capillary outlet, the number concentration of particles is assumed to be 1, and the number concentration of particles in other regions is expressed as a non-dimensional parameter.

With the above boundary conditions, the physical fields within the condenser segment can be represented as a mesh, and the temperature field, partial pressure field of supersaturated working medium vapor, and particle number concentration field can be solved using the finite difference method.

The temperature control equation of the condenser segment is discretized using the Taylor series, yielding the following equation:

$$u_{i,j}\frac{T_{i,j+1}-T_{i,j}}{\Delta z}=\alpha\frac{1}{r_{i,j}}\frac{\left(r_{i,j}-r_{i+1,j}\right)\frac{T_{i+1,j}-T_{i,j}}{\Delta r}-\left(r_{i-1,j}-r_{i,j}\right)\frac{T_{i-1,j}-T_{i,j}}{\Delta r}}{\Delta r} \tag{4.12}$$

Combine the above equation and simplify it to:

$$T_{i,j+1}=T_{i,j}+\frac{\alpha\Delta z}{\Delta r\cdot r_{i,j}\cdot u_{i,j}}\left(\left(r_{i,j}-r_{i+1,j}\right)\frac{T_{i+1,j}-T_{i,j}}{\Delta r}-\left(r_{i-1,j}-r_{i,j}\right)\frac{T_{i-1,j}-T_{i,j}}{\Delta r}\right) \tag{4.13}$$

$$T_{i,j+1}=T_{i,j}+\frac{\alpha\Delta z}{\Delta r\cdot r_{i,j}\cdot u_{i,j}}\left(\left(r_{i,j}-r_{i+1,j}\right)\frac{T_{i+1,j}-T_{i,j}}{\Delta r}-\left(r_{i-1,j}-r_{i,j}\right)\frac{T_{i-1,j}-T_{i,j}}{\Delta r}\right) \tag{4.19}$$

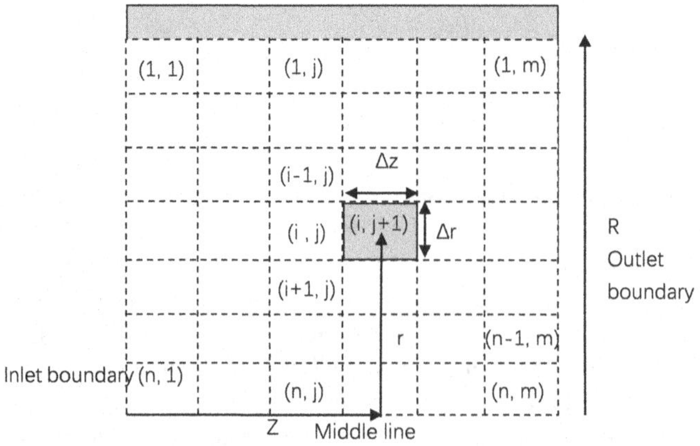

FIGURE 4.8 Differential computing grid.

In equation (4.13), $T_{i,j+1}$ can be determined based on the parameters at three points: $T_{i,j}, T_{i+1,j}, T_{i-1,j}$. By considering the boundary conditions, the temperature parameters of the j node can be calculated and solved successively along the Z-axis direction. Similarly, the particle number concentration field and the saturated working medium vapor partial pressure field in the condenser segment can be sequentially calculated in a similar manner.

4.3.3 Example of Solving Design Parameters of Condenser Segment

The mathematical simulation model of the condenser, especially for high-temperature CPC, can be constructed using MATLAB or other suitable platforms. By employing the finite difference method, the temperature field, saturated vapor pressure field, supersaturation field, and Kelvin equilibrium diameter distribution in the steady state of the condenser segment can be obtained.

The physical fields of the condenser segment model, as shown in Figures 4.9 and 4.10, are obtained under specific conditions, with the saturation segment being maintained at 300°C and the condenser segment at 250°C. The simulation provides valuable insights into the behavior and characteristics of the condenser under these operating parameters.

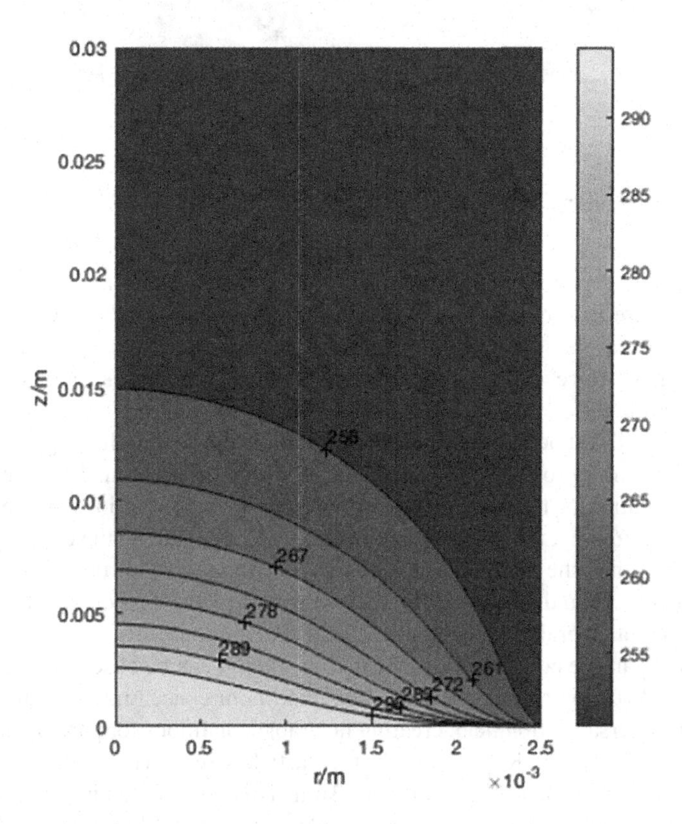

FIGURE 4.9 Temperature field distribution.

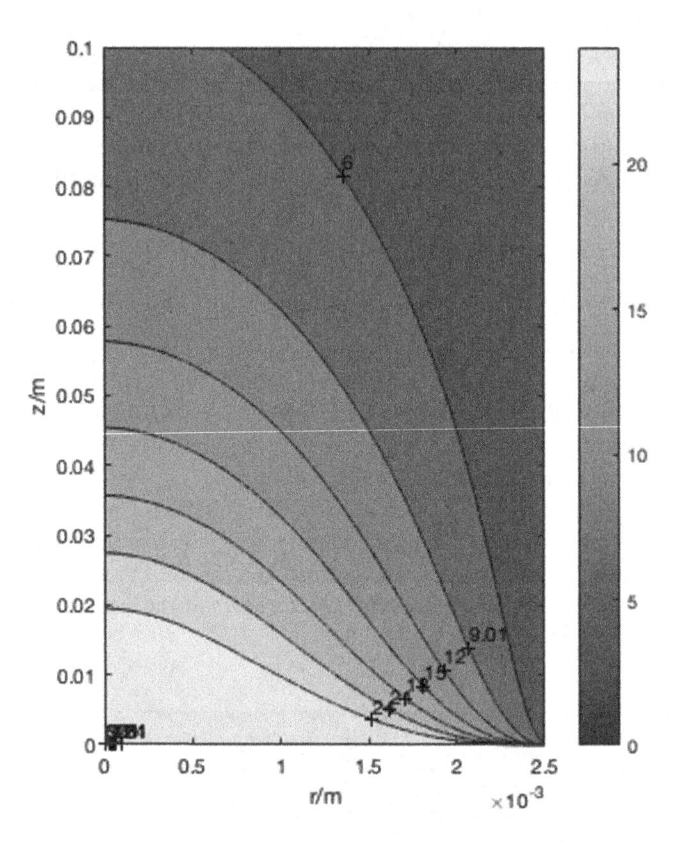

FIGURE 4.10 Partial pressure field distribution of steam working medium.

The temperature field plays a crucial role in influencing the distribution of partial pressure and supersaturation of steam in the saturated working medium. Analyzing the temperature field reveals that a significant temperature gradient exists only in a small region near the inlet of the condenser segment, approximately within a 10 mm distance range. However, the rest of the temperature field remains nearly equal to the wall temperature.

Examining the supersaturation field distribution shows that the largest saturation area is in proximity to the cooling segment in the central region. The three-dimensional distribution of the supersaturated field provides a more intuitive representation, highlighting that the maximum temperature gradient from the condenser to the exit is greater than the main distribution area of saturation. This observation implies that a higher thermal diffusion coefficient compared to the mass diffusion coefficient contributes to a favorable supersaturated field distribution.

From the particle concentration distribution field, it can be observed that particles show weak diffusion toward the wall. The maximum concentration aligns with the maximum supersaturation field, creating favorable conditions for particle nucleation.

Lastly, analyzing the Kelvin equilibrium particle size distribution reveals that the minimum activated particle size limit is 5.3 nm. This information is critical in understanding the behavior of particles in the condenser segment (Figures 4.11–4.14).

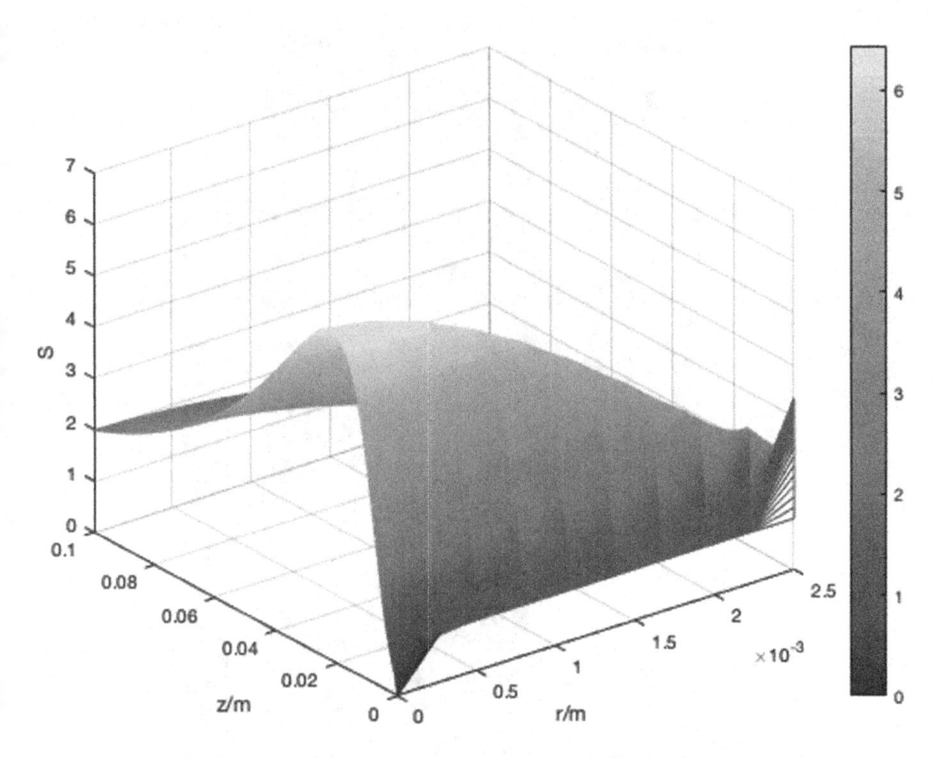

FIGURE 4.11 Three-dimensional supersaturation distribution.

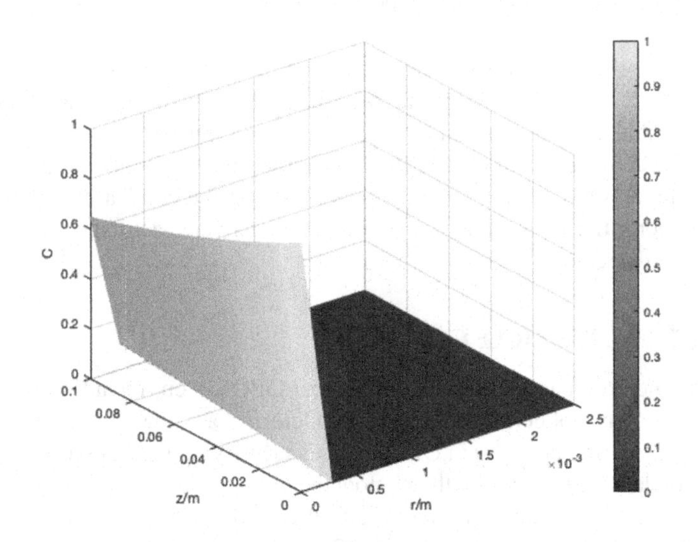

FIGURE 4.12 Field particle concentration field (particle size 10 nm, standardized).

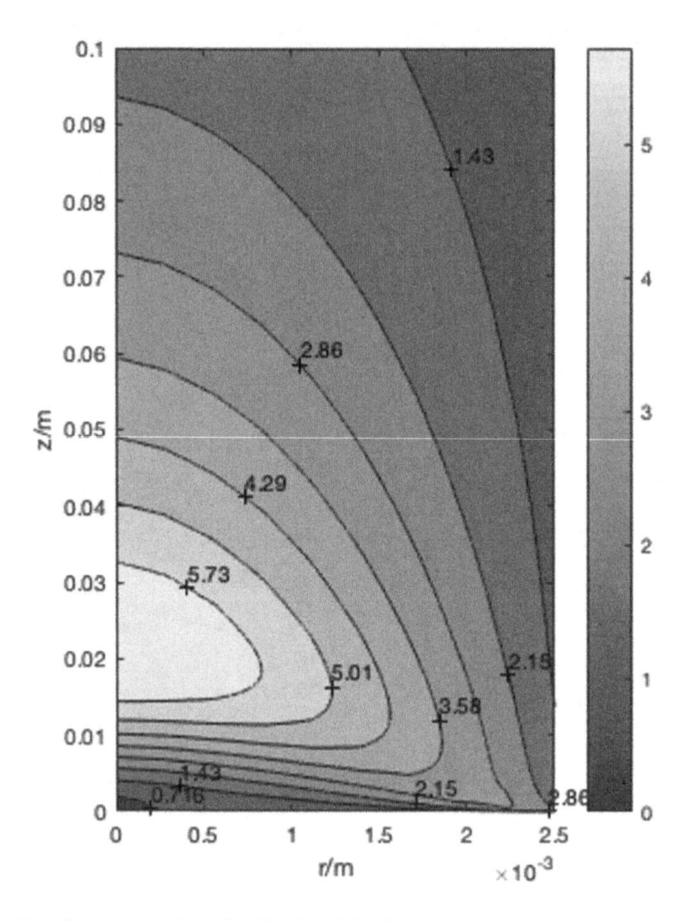

FIGURE 4.13 Supersaturation distribution field diagram.

Figure 4.15 illustrates the condenser growth process of particles with various sizes, namely 5, 8, 10, 15, and 23 nm, within the condenser segment. Notably, the particles exhibit rapid growth and can reach the micron level quickly during this process. Therefore, it is reasonable to assume that the particles can grow to sizes detectable by an OPC once they are activated within the condenser segment. This finding supports the notion that the condenser effectively facilitates the growth of particles to a detectable range.

4.4 OPTICAL PARTICLE COUNTER

OPC is another key part of the CPC. A typical OPC system includes laser system, laser focusing lens, collecting lens, and photodetector, as shown in Figure 4.16. The principle of detecting particles in OPC involves identifying the scattering light emitted from condensed particles by the photodetector.

The procedure to count particles in an OPC is described here. Firstly, a well-condensed aerosol particle goes into an OPC system and passes a light beam injected from the laser and laser-focusing lens systems. Then, the particle scatters the light as

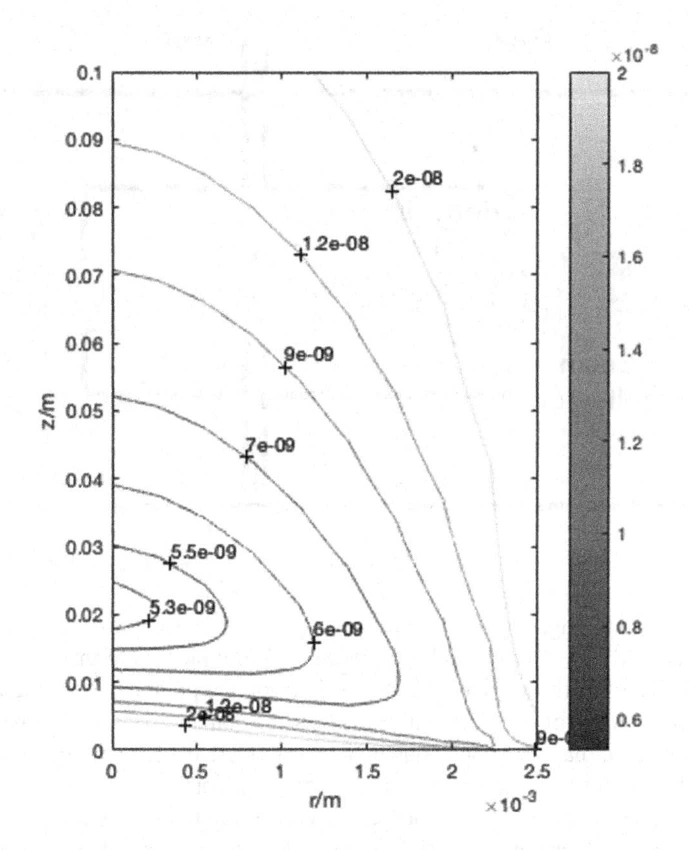

FIGURE 4.14 Kelvin equilibrium particle size distribution.

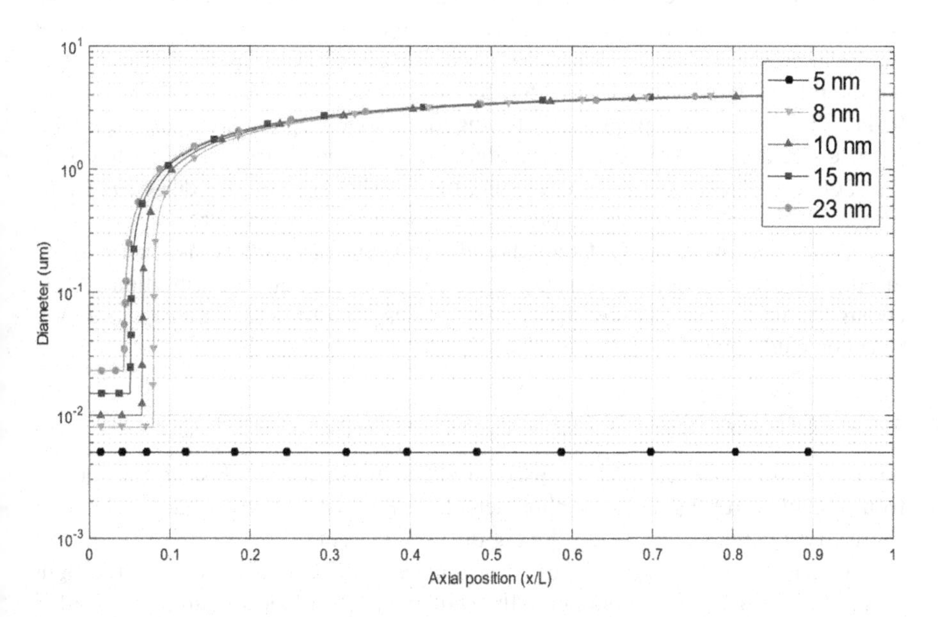

FIGURE 4.15 Model of particle condenser growth process.

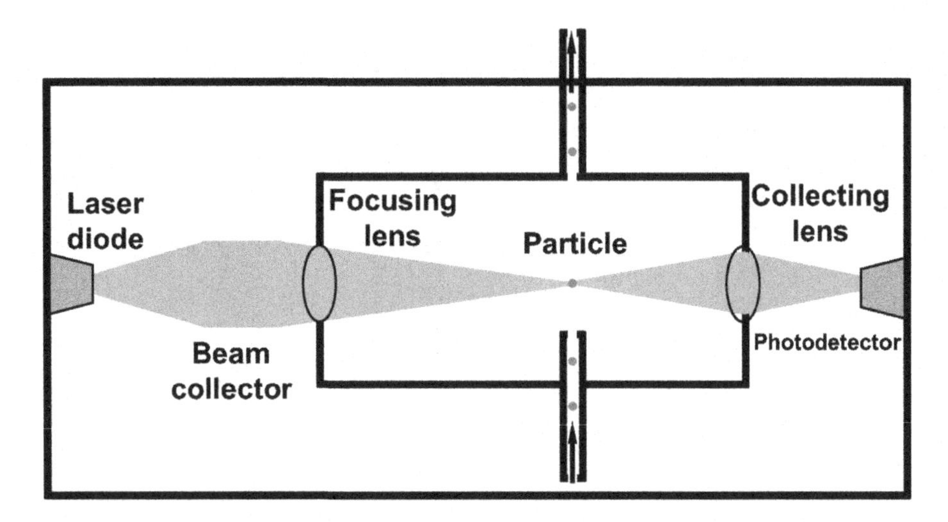

FIGURE 4.16 Schematic of the optical particle counter in condensation particle counter.

passing through the light beam. Before being detected by the low-noise photodetector, the scattered light is collected by a set of collecting lenses to intensify the light signal. Finally, after the scattered light irradiates on the photodetector, the light signal is converted into the electrical pulse to allow the count of particles by the counter. It should be noted that the optics housing is maintained at a higher temperature than the heated saturator to avoid condensation on the lens sets.

Besides the system described above, a reference photodiode is used to maintain a constant laser power output so that the particle signal can be identified. The light signal emitted from the particle is usually very weak than the reference light signal, so it should be intensified before comparing it to the reference value. For instance, the electrical pulse generated by a particle with a diameter of 0.5 μm is usually 100 mv, but the reference pulse is around 2.5V–5V so the electrical pulse should be intensified to 25–50 times its raw value. By tuning the voltage magnification, the minimum particle size detected by OPC can be adjusted. Since most particles exhausted from the cooled condenser are larger than 1μm, the threshold of OPC for detecting a particle can be set to 1μm to allow a precious and accurate measurement.

At low concentrations, OPC counts individual pulses produced as each particle (droplet) passes through the sensing zone. For high particle concentrations, OPC counts the total light scattered from the particles used to determine concentration based on calibration.

4.4.1 Light Source

An intense light source must be used to allow the detection of scattering pulses from a single aerosol particle by photodetectors. This subsection introduces the light source used in the history of developing the OPC system.

Halogen lamps were used in the early stages of OPC manufacture as the light source before lasers were commercially available [24]. A halogen lamp (also called tungsten halogen, quartz-halogen, and quartz iodine lamp) is an incandescent lamp consisting of a tungsten filament sealed in a compact transparent envelope that is filled with a mixture of inert gas and a small amount of a halogen, such as iodine or bromine. The combination of the halogen gas and the tungsten filament produces a halogen-cycle chemical reaction, which redeposits evaporated tungsten on the filament, increasing its life and maintaining the clarity of the envelope. This allows the filament to operate at a higher temperature than a standard incandescent lamp of similar power and operating life; this also produces light with higher luminous efficacy and color temperature. The small size of halogen lamps permits their use in compact optical systems for projectors and illumination. Halogen lamps are not favored by current OPC manufacturers since the light intensity is much weaker than lasers or state-of-the-art LEDs.

Lasers are widely used in OPC systems today due to their high power and excellent temporal coherence. Lasers are distinguished from other light sources by their coherence. Spatial (or transverse) coherence is typically expressed by the output being a narrow beam, which is diffraction-limited. Laser beams can be focused on very tiny spots, achieving a very high irradiance, or they can have very low divergence in order to concentrate their power at a great distance. Temporal (or longitudinal) coherence implies a polarized wave at a single frequency whose phase is correlated over a relatively great distance (the coherence length) along the beam [28]. A beam produced by a thermal or other incoherent light source has an instantaneous amplitude and phase that vary randomly with respect to time and position, thus having a short coherence length.

Lasers are characterized according to their wavelength in a vacuum. Most "single wavelength" lasers actually produce radiation in several modes with slightly different wavelengths. Although temporal coherence implies some degree of monochromaticity, there are lasers that emit a broad spectrum of light or emit different wavelengths of light simultaneously. Some lasers are not single spatial modes and have light beams that diverge more than is required by the diffraction limit. All such devices are classified as "lasers" based on their method of producing light by stimulated emission. Lasers are employed where light of the required spatial or temporal coherence cannot be produced using simpler technologies.

A light-emitting diode (LED) is used as an optional light source due to the breakthrough of white LED technology. Differing from early single-color LEDs, new high-power white-light LEDs can provide continuous, focused, and bright light. LEDs are a semiconductor light source that emits light when current flows through it. Electrons in the semiconductor recombine with electron holes, releasing energy in the form of photons (Energy packets). The color of the light (corresponding to the energy of the photons) is determined by the energy required for electrons to cross the band gap of the semiconductor [22]. White light is obtained by using multiple semiconductors or a layer of light-emitting phosphor on the semiconductor device [17].

Appearing as practical electronic components in 1962, the earliest LEDs emitted low-intensity infrared (IR) light [24]. Infrared LEDs are used in remote-control circuits, such as those used with a wide variety of consumer electronics. The first visible-light LEDs were of low intensity and limited to red.

LEDs have many advantages over incandescent light sources, including lower power consumption, longer lifetime, improved physical robustness, smaller size,

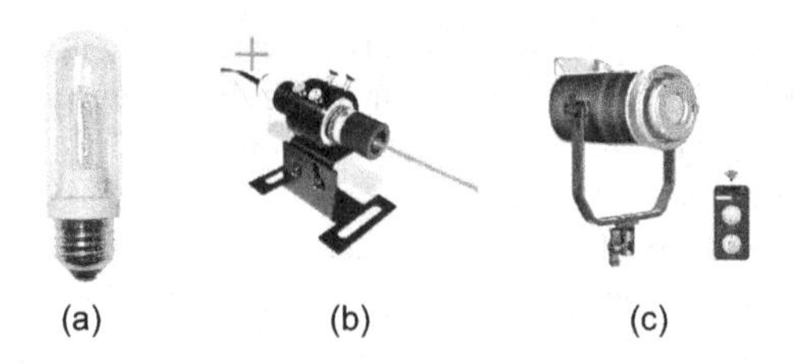

FIGURE 4.17 Light sources used in the OPC: (a) Halogen lamp, (b) Laser, and (c) LED.

and faster switching. In exchange for these generally favorable attributes, disadvantages of LEDs include electrical limitations to low voltage and generally to DC (not AC) power, inability to provide steady illumination from a pulsing DC or AC electrical supply source, and lower maximum operating temperature and storage temperature. In contrast to LEDs, incandescent lamps can be made to intrinsically run at virtually any supply voltage, can utilize either AC or DC current interchangeably, and will provide steady illumination when powered by AC or pulsing DC even at a frequency as low as 50 Hz. As a result, LED is a promising light source for OPC systems (Figure 4.17).

4.4.2 Lens

A lens is a transmissive optical device that focuses or disperses a light beam by means of refraction. A simple lens consists of a single piece of transparent material, while a compound lens consists of several simple lenses (elements), usually arranged along a common axis. Lenses are made from materials such as glass or plastic, and are ground and polished, or molded to a desired shape. Devices that similarly focus or disperse waves and radiation other than visible light are also called lenses, such as microwave lenses, electron lenses, acoustic lenses, or explosive lenses. In the optical system of the CPC, several different types of lenses are used to form a thin light sheet to illuminate particles or intensify scattering light signals.

Lenses can be classified into convex lenses and concave lenses. Convex lenses are thick in the middle and thinner at the edges. A concave lens is flat in the middle and thicker at the edges. A convex lens is also known as a converging lenses, as the light rays bend inward and converge at a point which is known as the focal length. On the other hand, the concave lens is also known as a diverging lens because it bends the parallel light rays outward and diverges them at the focal point.

A more comprehensive classification of lenses is biconvex lens, plano-convex lens, positive meniscus lens, negative meniscus lens, plano-concave lens, and biconcave lens (see Figure 4.18). Simple lenses are different from compound lenses based on their surface curvature. It should be noted that Fourier lenses are also used in CPC [20].

Lenses always introduce some degree of distortion or aberration that makes the image an imperfect replica of the object. Lens aberration also greatly influences the measurement accuracy of CPC by affecting the light sheet thickness. Several types of aberration affect image quality, including spherical aberration, coma, and chromatic aberration.

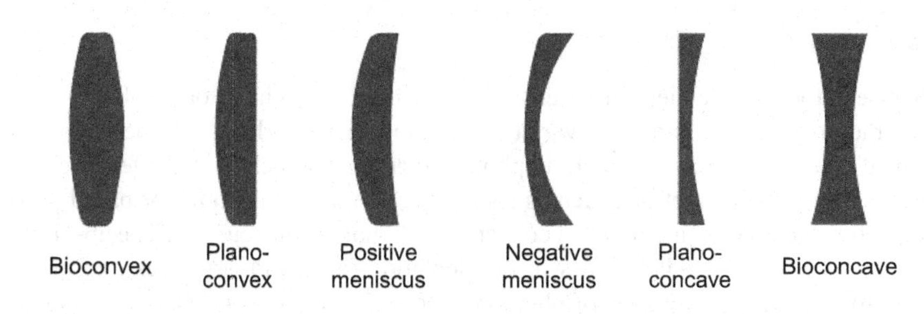

FIGURE 4.18 Typical lenses.

Spherical aberration occurs because spherical surfaces are not the ideal shape for a lens but are by far the simplest shape to which glass can be ground and polished and so are often used. Spherical aberration causes beams parallel to, but distant from, the lens axis to be focused in a slightly different place than beams close to the axis. This manifests itself as a blurring of the image. Spherical aberration can be minimized with normal lens shapes by carefully choosing the surface curvatures for a particular application. For instance, a plano-convex lens, which is used to focus a collimated beam, produces a sharper focal spot when used with the convex side toward the beam source.

Coma, or comatic aberration, derives its name from the comet-like appearance of the aberrated image. A coma occurs when an object of the optical axis of the lens is imaged. Rays that pass through the center of a lens of focal length f are focused at a point with distance $f \tan(\theta)$ from the axis. Rays passing through the outer margins of the lens are focused at different points, either further from the axis (positive coma) or closer to the axis (negative coma). In general, a bundle of parallel rays passing through the lens at a fixed distance from the center of the lens is focused on a ring-shaped image in the focal plane, known as a comatic circle. The sum of all these circles results in a V-shaped or comet-like flare. As with spherical aberration, coma can be minimized (and in some cases eliminated) by choosing the curvature of the two lens surfaces to match the application. Lenses in which both spherical aberration and coma are minimized are called best-form lenses.

Chromatic aberration is caused by the dispersion of the lens material—the variation of its refractive index n, with the wavelength of light. Since, from the formulae above, f is dependent upon n, it follows that light of different wavelengths is focused on different positions. The chromatic aberration of a lens is seen as fringes of color around the image. It can be minimized by using an achromatic doublet (or achromat) in which two materials with differing dispersion are bonded together to form a single lens. This reduces the amount of chromatic aberration over a certain range of wavelengths, though it does not produce perfect correction. The use of achromats was an important step in the development of the optical microscope. An apochromat is a lens or lens system with even better chromatic aberration correction, combined with improved spherical aberration correction. Apochromats are much more expensive than achromats. Different lens materials may also be used to minimize chromatic aberration, such as specialized coatings or lenses made from the crystal fluorite. This naturally occurring substance has the highest known Abbe number, indicating that the material has low dispersion.

4.4.3 Photodetector

Photodetectors, also called photosensors, are sensors of light or other electromagnetic radiation. There is a wide variety of photodetectors which may be classified by mechanism of detection, such as photoelectric or photochemical effects, or by various performance metrics, such as spectral response. Semiconductor-based photodetectors have a p–n junction that converts light photons into current. The absorbed photons make electron–hole pairs in the depletion region. Photodiodes and photo transistors are a few examples of photo detectors. Photodetectors may be classified by their mechanism for detection.

Photoemission detectors use the mechanism that photons cause electrons to transition from the conduction band of a material to free electrons in a vacuum or gas. Some typical photoemission detectors are Gaseous ionization detectors, Photomultiplier tubes, Phototubes, and Microchannel plate detectors. Gaseous ionization detectors are used in experimental particle physics to detect photons and particles with sufficient energy to ionize gas atoms or molecules. Electrons and ions generated by ionization cause a current flow that can be measured. Photomultiplier tubes contain a photocathode which emits electrons when illuminated. The electrons are then amplified by a chain of dynodes. Phototubes contain a photocathode which emits electrons when illuminated, such that the tube conducts a current proportional to the light intensity. Microchannel plate detectors use a porous glass substrate as a mechanism for multiplying electrons. They can be used in combination with a photocathode like the photomultiplier described above, with the porous glass substrate acting as a dynode stage.

Thermal photodetectors employ the principle that photons cause electrons to transition to mid-gap states and then decay back to lower bands, inducing phonon generation and, thus, heat. Bolometers measure the power of incident electromagnetic radiation via the heating of a material with a temperature-dependent electrical resistance. A microbolometer is a specific type of bolometer used as a detector in a thermal camera. Cryogenic detectors are sufficiently sensitive to measure the energy of single X-ray, visible, and infrared photons. Pyroelectric detectors detect photons through the heat they generate and the subsequent voltage generated in pyroelectric materials. Thermopiles detect electromagnetic radiation through heat and then generate a voltage in thermocouples. Golay cells detect photons by the heat they generate in a gas-filled chamber, causing the gas to expand and deform a flexible membrane whose deflection is measured.

In polarization photodetectors, photons induce changes in the polarization states of suitable materials, which may lead to changes in the index of refraction or other polarization effects. The photorefractive effect is used in holographic data storage. Polarization-sensitive photodetectors use optically anisotropic materials to detect photons of a desired linear polarization.

As for photodetectors using weak interaction effects, photons induce secondary effects such as in photon drag detectors or gas pressure changes in Golay cells.

4.4.4 Forward Scattering and 90° Scattering Systems

The forward-scattering spectrometer probe (FSSP) models (PMS) are ground-based or aircraft-mountable probes that size particles based on the intensity of forward-scattered light as they pass through a laser-illuminated sensing volume. The newer model FSSP-300 provides better sensitivity (down to 0.3 mm) and higher resolution (31 channels) over its range (0.3–20 mm) than the mechanically identical FSSP-100 (15 channels over several size ranges, such as 0.5–8.0 and 5.0–95 mm). The velocity operating range for these instruments is from about 10–125 m/s. The system has been extensively used in atmospheric aerosol studies.

The operating principles for the FSSP-100 have been described by and for the FSSP-300. A patented dual-detector arrangement is used to size only those particles passing through a prescribed sampling volume. Briefly, the sampling volume between two probe tips is illuminated by a HeNe laser from one of the tips. When a particle enters the volume, it scatters light, which is collected by optics located in the other probe tip. While a dump spot blocks the main beam, the forward-scattered light enters a beam-splitting prism and is focused on two photodetectors. The signal photodetector is unmasked and reports an intensity maximum used to size the particle, while the annulus detector is masked to eliminate light from in-focus, centered particles. A comparison between the two signals for each particle is used as an acceptance criterion: particles passing far from the focal plane scatter a larger proportion of light into the annular detector and are rejected. A transit time test is also performed to eliminate particles traversing the beam near an edge. This test can bias the size distribution, particularly for broad distributions with size-dependent particle velocities [31].

Several authors have reviewed the optical and electronic limitations of the FSSP technique [32,33]. Issues addressed include sample volume, sizing, and counting uncertainties. Concentration measurement uncertainties can be quite large [34], although correction algorithms can be applied to improve accuracy. Size calibration studies using latex and glass spheres and water droplets have been reported. Investigators have shown that experimental FSSP calibration curves are in reasonably good agreement with Mie calculations if particle refractive index, shape, and beam nonuniformities are considered (Figure 4.19).

A description of an in situ SPC based on white light illumination and detection at a 90° scattering angle was given by Umhauer [35]. The choice of white light illumination is intended to maximize the monotonicity of the scattering intensity versus diameter response curve, and to reduce (though not eliminate) index of refraction effects. These white light systems are well suited to filter efficiency testing, especially at high

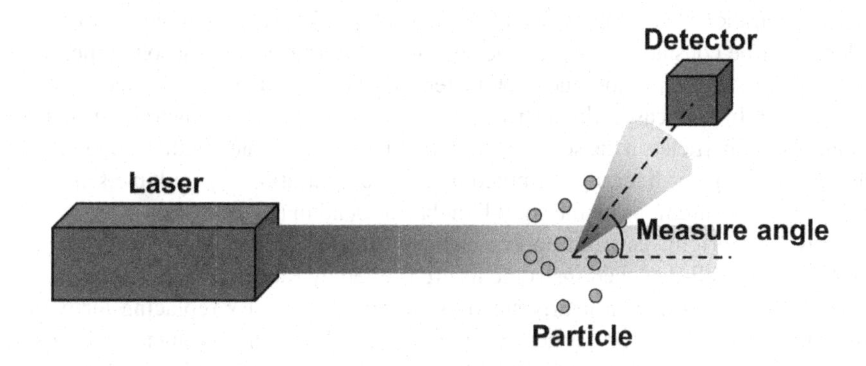

FIGURE 4.19 An example of a 90° scattering system.

pressures or temperatures, and have also been widely used in pharmaceutical spray sizing. The velocity operating range is typically from 0.1 to 10 m/s, although particle velocities are not measured. Commercial systems based on the Umhauer design have been marketed as the HC series particle sizers by PLY (now discontinued) and as the Model PCS-2000 series (PAS). Both manufacturers offered several models, differing in optical geometry and, hence, in nominal size and concentration ranges.

Depending on the configuration of the PCS-2000 system (PAL), the particle size range can be chosen between 0.15 and 100 mm for particle concentrations up to 10^6 particles/cm^3. PCS-2000 systems offer two independent photomultipliers to minimize border zone errors (i.e., trajectory ambiguity). Only a few studies of the PCS-2000 system have appeared in the literature. The performance of PCS-2000 systems is expected to be generally similar to the HC series (PLY) based on their shared ancestry [35]

4.4.5 Single Particle Counters and Multiple Particle Counters

The particle counter in CPC measures the size and concentration of aerosol particles in a limited size range by means of light scattering by single particles. For this purpose, a stream of aerosol is drawn through a condensed light beam. Light flashes scattered from single particles are received by a photodetector and converted into electrical pulses. From the count rate of the pulses, the number concentration, and the pulse height, the size of the particles is derived. The light power that an individual particle scatters is a function of its size, refractive index, and shape. Particle sizing based on this principle has been known for more than a century. The invention of the laser has allowed the successful replacement of white light illumination by coherent and monochromatic laser light. Important characteristic features of an OPC are its permissible range of number concentration, its sampling flow rate, its sensitivity (lower detection limit), and its size-measurement accuracy. The requirements for an OPC change for different kinds of applications. Thus, the specifications of an instrument have to be adjusted to the specific measurement problem. Optical particle counters that cover these various kinds of applications are offered, for instance, by CLI, PAC, and PMS.

Multiple particle detection instruments, such as photometers or nephelometers, based on scattered-light intensity are useful for concentration measurements of aerosols if certain requirements are met. For the determination of concentration ratios or relative concentrations, the composition of the aerosol (particle size distribution, refractive index) must be constant during a series of experimental runs. For absolute measurements of mass concentration, the photometer must be calibrated with the aerosol to be investigated. In both cases, the instrument must be operated in its linear range, where the number of particles in the sensing volume is linearly correlated with the photometer signal. This range of linearity is limited at high concentrations by multiple scattering and at low concentrations by the stray-light background in the chamber.

Stray light originates from optical elements such as lenses, glass windows, and optical stops. Well-designed instruments are limited by Rayleigh scattering from the air molecules, resulting in nearly steady-state noise levels. By replacing air with a gas of known scattering properties, it is possible to calibrate a photometer in terms of scattering cross-section. A measure of the stray-light background is the photometer response in the presence of particle-free air. In all cases where the analog signal is subjected to electronic data processing, the photometer should be adjusted to a zero response for particle-free air.

From the lower detection limit to the onset of multiple scattering, the linearity range of a typical photometer can span at least three orders of magnitude in number concentration. Aerosol photometers, in combination with inert aerosols, are currently applied in filter testing and aerosol medicine. In general, instruments based on light scattering are much more sensitive than light extinction systems.

4.5 CALIBRATION TECHNOLOGY AND ANALYSIS OF ERROR SOURCE IN CPC

4.5.1 Calibration Technology of CPC

CPC perform measurements of particulate matter number concentration by counting the electrical pulse signals generated by the particles. The calculation formula for this is shown in equation (4.14):

$$C = N/tQ \tag{4.14}$$

Where C is the measured particle number concentration (particles/cm^3); N is the count of voltage pulse signals; t is the sampling time (s); Q is the sampling flow (cm^3/s).

The calibration of CPC is moving toward the international trend of focusing on small particles and low concentrations. In the Euro VI emission standard, CPC is designated as the preferred instrument for measuring particle number concentration, making research on calibration technology highly significant. The earliest CPC calibration method can be traced back to Aitken's manual calculation method and the pipe bridge calibration device invented by Nolan and Pollak. There are currently three mainstream CPC calibration methods: the microscopic particle counting method, the aerosol electrometer method, and the standard CPC method. The microscopic particle counting method involves quantitatively collecting the flowing aerosols on a filter and using a scanning electron microscope to count the number of particles collected. However, this calibration method is not commonly used due to factors such as large measurement uncertainty and high operational difficulty resulting from the complexity of variables in the calibration process [26]. As a result, our main focus will be on the aerosol electrometer method and the standard CPC method. Additionally, we will introduce a new CPC calibration tool, the ink-jet aerosol generator.

The aerosol electrometer method, being the first CPC calibration method proposed and still used today, represents the highest internationally recognized standard for particle counting. The measurement results obtained from this method can be traced back to the measurement of electric current. The principle can be described as follows: when aerosol particles with a single charge enter the sampling port, they are captured by the high-efficiency filter cartridge located within the Faraday cup. Due to the space charge effect, the charge on the surface of particles is released in the Faraday cup, generating a measurable electric current. The particle concentration can be calculated based on the current and the sampling flow value (see Figure 4.20).

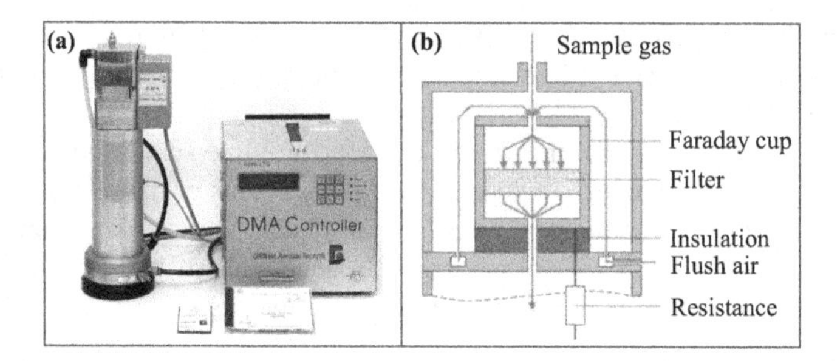

FIGURE 4.20 Electricity measurement material diagram (a) and schematic diagram (b).

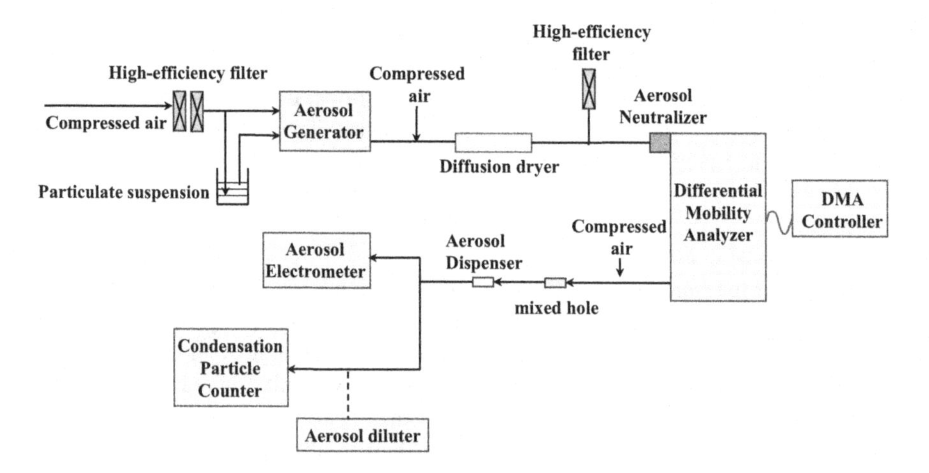

FIGURE 4.21 CPC calibration device (aerosol electrometer method).

The calibration method of CPC using aerosol electrometer was first proposed in reference [36], and the calibration process and device can be represented by Figure 4.21. In this method, an aerosol generator is used to produce monodisperse standard particles, which then pass through a diffusion dryer to remove water. The particles subsequently enter the aerosol neutralizer to achieve Boltzmann charge balance. The monodisperse aerosol particles of a specific size reaches a dynamic equilibrium in terms of particle concentration at the outlet of the aerosol dispenser, achieved through techniques such as dilution and replenishment of clean air, and flows through the CPC to be calibrated and the calibrated aerosol electrometer at a specific flow rate. Finally, the calibration of CPC is completed by comparing the results of aerosol electrometer and CPC. The linear relationship between Germany GRIMM 5705 static electricity meter and CPC is represented by the equation $y = 0.998x$ ($R^2 = 0.9999$).

A previous study calibrated the TSI 3025 CPC using the above method, and the sodium chloride and silver particles in the size range of 2–20 nm were used as the standard aerosols. In order to minimize particle loss during the experiment, a higher aerosol flow (1.5 L/min) and a shorter aerosol tube (less than 10 cm) were employed.

Collins et al. proposed a calibration method for high-particle concentration, utilizing the Langbo W function to solve the CPC theory compliance rate. This method is based on the pine process, which is used to correct the coincidence in the CPC. When comparing it with the standard aerosol electrometer method, the average error for TSI 3760A CPC at a concentration of $8 \times 104 cm^{-3}$ is only 4%. Chung et al. conducted a comparison between the effects of using spray aerosol generators and cigarette aerosol generators for particle source calibration on nanoparticle concentration and linearity. The study found that the CPC has a higher counting efficiency for the spray aerosol generator, which indicates that the spray aerosol generator can generate larger particles compared to the cigarette aerosol generator [27].

The standard CPC method for CPC calibration is conducted by comparing it with a calibrated CPC. The calibration device used is basically the same as shown in Figure 4.21, with the only difference being that the aerosol electrometer is replaced with a standard CPC. Rose et al. conducted calibration experiments and assessed the measurement uncertainty for a continuous flowing cloud condensation nuclei particle counter [28]. They determined the effective water vapor supersaturation (SS) in the range of 0.05%–1.4% by measuring the dry particle activation diameter and calculating it using the Kohler model. The results show that, at high SS values, the relative standard deviation between the measured and the theoretically calculated results was less than 10%. However, when $SS \leq 0.1\%$, the relative standard deviation exceeded 40%. Therefore, to achieve high-precision, cloud condensation nuclei particle counters require more precise experimental calibration at low SS values. In the calibration study of cloud condensation nuclei particle counters conducted by Kuwata et al., it was mentioned that when the Pitzer model was used to calculate the water activity, the calculated values for both ammonium sulfate and sodium chloride particles were within 2% relative uncertainty, indicating that theoretical calculations using the Pitzer model gave the most accurate SS values [29]. Giechaskiel et al. evaluated the effect of aerosol chemical composition on CPC calibration results [30]. By comparing the counting efficiency curves for different material particles, they found that the measurement uncertainty was 0.1 when the counting efficiency exceeded 0.9. Furthermore, it was observed that the counting efficiency at a particle size of 23 nm is greatly affected by the chemical composition of the particles. These findings align closely with the observations made during the calibration of the engine exhaust condensation nuclei particle counter.

Fletcher et al. conducted a comprehensive experimental comparison of three calibration techniques for CPC: microscopic particle counting method, aerosol electrometer method, and standard CPC method [26]. The experiments revealed that the measurement results obtained using the aerosol electrometer method or the standard CPC method can be affected by polymers present in the particles emitted by aerosol generator. This effect can be minimized by using a spray-type aerosol generator. The uncertainty associated with aerosol electrometer measurement and standard CPC measurement is relatively similar. However, when the particle concentration exceeds $5{,}000 cm^{-3}$, the uncertainty of the standard CPC method is slightly higher, at approximately 5%. Within the concentration range of 3,000 to $5{,}000 cm^{-3}$, the measurement uncertainties of aerosol electrometer measurements and scanning electron microscope measurements are comparable. Nevertheless, the aerosol electrometer exhibits lower uncertainty when the particle concentration exceeds 5,000 cm^{-3}, while scanning electron microscope demonstrates lower uncertainty at the particle concentration below 3,000 cm^{-3}, especially at very low concentrations. The exponential increase in measurement uncertainty is attributed to the fact that at low concentrations, the total charge number is so minuscule that it falls below the lower detection limit of the aerosol electrometer. In ten separate aerosol test experiments conducted at different concentrations, the concentration detection results obtained through scanning electron microscopy were consistently lower than those obtained using the other two methods. Additionally, the scanning electron microscope exhibited a measurement uncertainty of approximately 3% for all concentrations.

The aforementioned methods involve using an aerosol generator to produce aerosol particles and calibrating the CPC alongside a comparison instrument to measure and compare the aerosol airflow simultaneously. Iida et al. developed an ink-jet aerosol generator capable of generating highly monodisperse solid or liquid aerosol particles in the size range of 0.3–20 μm, with precise specific aerosol generation rates. Utilizing this generator for calibrating the CPC eliminates the need for intermediate steps, such as the aerosol electromigration separator in the screening process and the standard instrument for measurement. Instead, it allows for a direct comparison of nominal data between the ink-jet aerosol generator and the CPC to be calibrated. Kesavan et al. tested the accuracy of ink-jet aerosol generators with a particle size analyzer. The relative detection efficiency, which represents the ratio of the number of detected particles to the number of particles nominally generated by the ink-jet aerosol generator, was found to be approximately 99.3% for particles in the size range of 2–13.3 μm. However, there was a slight drop of about 2% in the relative detection efficiency for particles in the range of 0.95–2 μm.

4.5.2 Analysis of Error Sources in CPC

1. Environmental error

   This error is caused by the influence of various factors such as temperature, humidity, air pressure, electromagnetic fields, mechanical vibrations, sound, light, radioactivity, etc.

2. Personal error

   This error arises from limitations in human sense organs and motor skills. They occur in measurements that rely on human judgment through sight and hearing, as well as in manual adjustments. Examples include reading the wrong scale or misinterpreting readings.

3. Instrument error

   Instrument errors stem from imperfections in the electrical or mechanical properties of the instrument itself and its accessories [37]. Examples of instrument errors include errors caused by the standard resistance in the bridge and the probe line of the oscilloscope. Additionally, instrument errors can result from zero offsets, scale inaccuracies, and nonlinearity.

4. Operation error

   Operation error is caused by improper installation, adjustment, arrangement, and use of the instrument. For instance, placing the instrument horizontally instead of vertically as required, poor grounding of the instrument, using excessively long test leads leading to loss or impedance mismatch, and performing measurements without following the proper preheating, adjustment, and calibration procedures.

5. Method error

   Method error occur due to inappropriate modification and simplification of classical measurement methods. These errors can result from the use of imperfect measurement methods and the lack of a rigorous theoretical basis. Such errors encompass all factors not accounted for in the expression of the measurement results and are also referred to as theoretical errors. For example, when using an ordinary multimeter to measure the voltage across a high-resistance resistor in a circuit, a shunt may form due to the low internal resistance of the multimeter's voltage block, leading to measurement errors.

4.5.3 Conclusion

1. Excessive concentration of particulate matter in the sampling airflow will reduce the supersaturation of the working liquid vapor, which will affect the CPC counting efficiency. The effect of excessive concentration on the CPC counting efficiency can be remedied simply by diluting the sampling airflow.
2. The counting efficiency of CPC is optimal when there is a flow rate difference between the sample gas flow rate in the capillary and the sheath gas flow rate in the supersaturated region.
3. Reducing the operating pressure within a certain range can improve the counting efficiency of the CPC.
4. The effect of humidity on the CPC detection efficiency is not absolute and depends on the hygroscopic properties of the aerosol particles and their contact angle with water vapor.
5. Particle loss can be minimized during calibration by using a larger aerosol flow or by shortening the aerosol airway.
6. The influence of the small amount of polymer particles generated by the aerosol generator on the calibration results should be minimized, and a spray-type aerosol generator is a good choice.
7. It should use the same test conditions as the measurement environment as much as possible for calibrating the CPC.

4.6 COMPARISON BETWEEN DIFFERENT COMMERCIAL CPCS

There are several commercial CPCs in the current market, and they are widely used in different field, such as environmental research, occupational health and safety, atmospheric studies, and aerosol science, for measuring the characteristics of particle matters coming from different resources. In this section, some of the popular commercial CPCs produced by well-known manufacturers are listed and compared as follows.

1. TSI Incorporated is a global company designing and producing high-precision measurement instruments, and they have some popular CPCs, such as Model 3752 and Model 3756 [38].
   - Pros:
     a. High accuracy: TSI CPCs are known for their precise particle counting capabilities.
     b. Wide dynamic range: They can measure particles across a broad size range (typically 0.002–10 μm).
     c. Advanced detection technology: TSI utilizes advanced condensation growth or optical detection techniques for accurate results.
     d. Real-time measurement: TSI CPCs provide real-time or near-real-time monitoring of particle concentrations.
     e. User-friendly interface: TSI instruments often have intuitive interfaces and software for easy operation and data analysis.
   - Cons:
     a. TSI CPCs have a higher price point than some other brands.
     b. Complexity: Some models have a more intricate setup and operation process.
     c. Technically, it relies on traditional condensation mechanisms.
2. AVL is a global company that manufactures advanced testing and measurement equipment for the automotive industry, particularly in the field of powertrain development and testing, and they also have particle counters [39].
   - Pros:
     a. Wide dynamic range: the particle size detected by AVL can be up to 200 nm
     b. Robust construction: Robust and precise dilution system without particle size dependency
     c. High accuracy and precision: AVL provides high accuracy in measuring particles due to immediate dilution at the sampling position
     d. Excellent consistency: AVL provides very robust measurement for detecting particles over time, even with varying environmental conditions
     e. Compact design: Easily transported with low space requirements
   - Cons:
     a. Limited sensitivity for particles smaller than 0.1 μm
     b. AVL provides very limited CPC options compared to other companies
3. Grimm Aerosol Technik is a well-known company specializing in the development and production of aerosol measurement instruments, and they have some CPC models, such as 5403, 5416, and 5430/5431 [40].
   - Pros:
     a. High accuracy: Grimm CPCs are designed to provide accurate particle counting.
     b. Wide dynamic range: They can detect particles across a wide size spectrum (4.5 nm to >3 μm).

c. Real-time monitoring: Grimm CPCs offer real-time measurement capabilities for continuous monitoring.
d. Customizable options: Grimm provides customization options to adapt the instrument to specific needs.
e. Robust construction: Grimm instruments are known for their durable construction and long-term reliability.

- Cons:
  a. Limited model options: The range of available Grimm CPC models is limited.
  b. Poor performance in high-concentration aerosol environments

4. Kanomax is a leading manufacturer focused on precision measuring instruments and solutions with a specialization in the field of air quality monitoring and particle measurement, and they have several CPC models, such as scanning threshold particle counter model 9010, fast condensation particle counter model 3650 etc. [41].
   - Pros:
     a. High accuracy and sensitivity: Kanomax CPCs provide accurate and sensitive particle measurements.
     b. Wide dynamic range: They can detect particles across a wide size spectrum (typically 10 nm to 10 μm).
     c. Real-time monitoring: Kanomax CPCs offer real-time measurement capabilities for continuous monitoring.
     d. Robust construction: Kanomax instruments are known for their durable construction and reliability.
     e. Lower cost: It also has a relatively low cost compared to other CPCs.
   - Cons:
     a. Kanomax CPC has a lower counting efficiency than some other models and may not be suitable for certain applications.
     b. Different Kanomax CPC models have variations in measurement range, flow rates, and other specifications.
5. Palas GmbH is a respected manufacturer of scientific instruments, including CPCs, and they have several models available in the market, such as ENVI-CPC50/100/200 [42].
   - Pros:
     a. Accuracy and Sensitivity: Palas CPCs are typically designed to provide accurate and sensitive measurements of particle concentrations.
     b. Wide Particle Size Range: Palas CPCs often offer a broad measurement range, allowing for the detection and counting of particles across a wide size spectrum (typically 10 nm to 100 μm).
     c. Real-Time Monitoring: Palas CPC provides real-time or near-real-time measurement capabilities, enabling continuous monitoring and observation of particle concentrations and size distributions.
     d. Robust Construction: Palas instruments are known for their durable construction, ensuring reliability and stability in measurement performance.

- Cons:
    a. the available options are specific to certain applications or research areas.
    b. Different Palas CPC models have variations in terms of measurement range, flow rates, and other specifications.

In sum, there are different types of commercial CPCs available in the current market, and different CPC models offer specific features, advantages, and limitations that provide options for different requirements and applications.

REFERENCES

[1] P. Kulkarni, P. A. Baron, and K. Willeke. *Aerosol Measurement: Principles, Techniques, and Applications*. John Wiley & Sons, 2011.
[2] J. Aitken. "I.-On the number of dust particles in the atmosphere." *Earth and Environmental Science Transactions of the Royal Society of Edinburgh* 35, no. 1 (1889): 1–19.
[3] N. Collings, K. Rongchai, and J. P. R. Symonds. "A condensation particle counter insensitive to volatile particles." *Journal of Aerosol Science* 73 (2014): 27–38.
[4] G. J. Sem. "Design and performance characteristics of three continuous-flow condensation particle counters: a summary." *Atmospheric Research* 62, no. 3–4 (2002): 267–294.
[5] J. Kangasluoma, S. Hering, D. Picard, etc. "Characterization of three new condensation particle counters for sub-3 nm particle detection during the Helsinki CPC workshop: the ADI versatile water CPC, TSI 3777 nano enhancer and boosted TSI 3010." Atmospheric Measurement Techniques 10, no. 6 (2017): 2271–2281.
[6] G. Mordas, M. Kulmala, T. Petäjä, etc. "Design and performance characteristics of a condensation particle counter UF-02proto." *Boreal Environment Research* 10, no. 6 (2005): 543.
[7] M. Hermann, B. Wehner, O. Bischof, etc. "Particle counting efficiencies of new TSI condensation particle counters." *Journal of Aerosol Science* 38, no. 6 (2007): 674–682.
[8] X. Zhang. "Development and investigation on a high temperature condensation particle counter", Master of Thesis, Beihang University, 2016.
[9] Y.G. Ma, "Numerical simulation and experimental study on condensation growth of high temperature condensation particle counter", Master of Thesis, Beihang University, 2017.
[10] D. Ye. "Performance verification of a wide-temperature condensation particle counter", Master of Thesis, Beihang University, 2018.
[11] M. R. Stolzenburg, and P. H. McMurry. "An ultrafine aerosol condensation nucleus counter." *Aerosol Science and Technology* 14, no. 1 (1991): 48–65.
[12] L. F. Chen, X. Zhang, C. Q. Zhang, M. Raza, and X. H. Li. "Experimental investigation of a condensation particle counter challenged by particles with varying wettability to working liquid." *Aerosol and Air Quality Research* 17 (2017): 2743–2750.
[13] M. Kupper, M. Kraft, A. Boies, and A. Bergmann. "High-temperature condensation particle counter using a systematically selected dedicated working fluid for automotive applications." *Aerosol Science and Technology* 54 (2020): 381–395.
[14] K. Iida, M. R. Stolzenburg, and P. H. McMurry. "Effect of working fluid on sub-2 nm particle detection with a laminar flow ultrafine condensation particle counter." *Aerosol Science and Technology* 43 (2009): 81–96.
[15] L. E. Magnusson, J. A. Koropchak, M. P. Anisimov, V. M. Poznjakovskiy, and J. F. de la Mora. "Correlations for vapor nucleating critical embryo parameters." *Journal of Physical and Chemical Reference Data* 32 (2003): 1387–1410.

[16] W. Wang. "Research on key technologies of on-line measurement of ultrafine particle number concentration of motor vehicle emission based on condensation particle counting", Doctor, University of Science and Technology of China, 2020. https://doi.org/10.27517/d.cnki.gzkju.2020.001144.

[17] W. Wang, X. Zhao, J. Zhang, Y. Yang, T. Yu, J. Bian, H. Gui, and J. Liu. "Design and evaluation of a condensation particle counter with high performance for single-particle counting." *Instrumentation Science & Technology* 48 (2020): 212–229. https://doi.org/10.1080/10739149.2019.1687516.

[18] S. Biswas, P. M. Fine, M. D. Geller, S. V. Hering, and C. Sioutas, "Performance evaluation of a recently developed water-based condensation particle counter." *Aerosol Science and Technology* 39 (2005): 419–427. https://doi.org/10.1080/027868290953173.

[19] S. V. Hering, and M. R. Stolzenburg. "A method for particle size amplification by water condensation in a laminar, thermally diffusive flow." *Aerosol Science and Technology* 39 (2005): 428–436. https://doi.org/10.1080/027868290953416.

[20] S. V. Hering, M. R. Stolzenburg, F. R. Quant, D. R. Oberreit, and P. B. Keady. "A laminar-flow, water-based condensation particle counter (WCPC)." *Aerosol Science and Technology* 39 (2005): 659–672. https://doi.org/10.1080/02786820500182123.

[21] F. J. Romay, A. M. Collins, W. D. Dick, L. Li, C. W. Fandrey, and B. Y. H. Liu. "Water-based single-flow mixing condensation particle counter." *Aerosol Science and Technology* 50 (2016): 1320–1326. https://doi.org/10.1080/02786826.2016.1222510.

[22] S. V. Hering, G. S. Lewis, S. R. Spielman, A. Eiguren-Fernandez, N. M. Kreisberg, C. A. Kuang, and Attoui, M. "Detection near 1-nm with a laminar-flow, water-based condensation particle counter." *Aerosol Science and Technology* 51 (2017): 354–362. https://doi.org/10.1080/02786826.2016.1262531.

[23] C. Zhang, X. Zhang, Y. Guo, D. Zhang, and L. Chen. *Explore the Influence Law of Counting Efficiency of High Temperature Condensation Core Particle Counter.* Internal Combustion Engine Technology, 2015.

[24] P. H. McMurry. "The history of condensation nucleus counters." *Aerosol Science and Technology* 33 (2000): 297–322. https://doi.org/10.1080/02786820050121512.

[25] C. M. Sorensen, J. Gebhart, T. J. O'Hern, and D. J. Rader. "Optical measurement techniques: fundamentals and applications." In: *Aerosol Measurement: Principles, Techniques, and Applications.* John Wiley & Sons, Inc., Hoboken, NJ, 2011: pp. 269–312. https://doi.org/10.1002/9781118001684.ch13.

[26] R. A. Fletcher, G. W. Mulholland, M. R. Winchester, R. L. King, and D. B. Klinedinst. "Calibration of a condensation particle counter using a NIST traceable method." *Aerosol Science and Technology* 43, no. 5 (2009): 425–441. https://doi.org/10.1080/02786820802716735.

[27] M. C. Chung, M. S. Kim, G. S. Sung, S. M. Kim, and J. W. Lee. "Comparison study on characteristics of nano-sized particle number distribution by using condensation particle counter calibrated with spray and soot type particle generation methods." *International Journal of Automotive Technology* 15, no. 6 (2014): 877–884. https://doi.org/10.1007/s12239-014-0092-9.

[28] D. Rose, S. S. Gunthe, E. Mikhailov, G. P. Frank, U. Dusek, M. O. Andreae, and U. Pöschl. "Calibration and measurement uncertainties of a continuous-flow cloud condensation nuclei counter (DMT-CCNC): CCN activation of ammonium sulfate and sodium chloride aerosol particles in theory and experiment." *Atmospheric Chemistry and Physics* 8, no. 5 (2008): 1153–1179. https://doi.org/10.5194/acp-8-1153-2008.

[29] M. Kuwata, and Y. Kondo. "Measurements of particle masses of inorganic salt particles for calibration of cloud condensation nuclei counters." *Atmospheric Chemistry and Physics* 9, no. 16 (2009): 5921–5932. https://doi.org/10.5194/acp-9-5921-2009.

[30] B. Giechaskiel, X. Wang, H. G. Horn, J. Spielvogel, C. Gerhart, J. Southgate, L. Jing, M. Kasper, Y. Drossinos, and A. Krasenbrink. "Calibration of condensation particle counters for legislated vehicle number emission measurements." *Aerosol Science and Technology* 43, no. 12 (2009): 1164–1173. https://doi.org/10.1080/02786820903242029.
[31] B. Baumgardrier, W. A. Cooper, and J. E. Dye. "Optical and electronic limitations of the forward-scattering spectrometer probe." In: *Liquid Particle Size Measurement Techniques: 2nd Volume*. ASTM International, 1990: p. 115.
[32] D. Baumgardner, J. E. Dye, B. W. Gandrud, and R. G. Knollenberg. "Interpretation of measurements made by the forward scattering spectrometer probe (FSSP-300) during the Airborne Arctic Stratospheric Expedition." *Journal of Geophysical Research: Atmospheres* 97 (1992): 8035–8046.
[33] M. Wendisch, A. Keil, and A. V Korolev. "FSSP characterization with monodisperse water droplets." *Journal of Atmospheric and Oceanic Technology* 13 (1996): 1152–1165.
[34] R. G. Knollenberg. "Techniques for probing cloud microstructure." In *Clouds, their Formation, Optical Properties, and Effects*, 1981: pp. 15–89.
[35] H. Umhauer. "Particle size distribution analysis by scattered light measurements using an optically defined measuring volume." *Journal of Aerosol Science* 14 (1983): 765–770.
[36] B. Liu and D. Pui. "On the performance of the electrical aerosol analyzer." *Journal of Aerosol Science* 6 (1975): 249–264.
[37] A. Terres, B. Giechaskiel, A. Nowak, and V. Ebert. "Calibration uncertainty of 23nm engine exhaust condensation particle counters with soot generators: A European automotive laboratory comparison." *Emission Control Science and Technology* 7 (2021): 124–136
[38] CondensationParticleCounters. https://tsi.com/products/particle-counters-and-detectors/condensation-particle-counters/
[39] AVL Particle Counter. https://www.avl.com/documents/10138//885965//AVL+Particle+Counter.pdf
[40] Grimm Aeros. https://www.directindustry.com/prod/grimm-aerosol-technik 69071.html.
[41] Kanomax FMT Prod. https://www.kanomaxfmt.com/product/
[42] Palas ENVI- CPC. https://www.palas.de/en/product/envicpcSystem

5 Application of Particle Condensation Counting Technology

Jingsha Xu, Zhirong Liang, Boxuan Cui, and Xuehuan Hu

5.1 INTRODUCTION

Condensation particle counter (CPC) is capable of measuring fine particles from stationary sources, mobile sources, and those existing in ambient atmosphere. The fine particles are with aerodynamic diameter of <1 μm, while the ultrafine particles are with aerodynamic diameter of <0.1 μm, which typically accounts for more than 80% of $PM_{2.5}$ by number [1]. These small particles have great impacts on environmental quality and human health, and they are prone to experience complex physio-chemical processes including adsorption, nucleation, condensation, growth, evaporation, deposition, and chemical reactions in the atmosphere. For instance, these particles can rapidly transform into droplets under supersaturation conditions and then become a potentially important contributor to cloud condensation, the process of which is very important to regional transport of heavy haze pollution and even global climate change [2]. Nowadays, although there are numerical-based air quality models that can quantify the comprehensive evolution of the particles, they cannot fully consider and precisely assess the growth of particles in the atmosphere. The key factors such as particulate nucleation and coagulation affecting the growth process of the particles at various conditions need to be well considered, where the CPC can be implemented for adequate measurements.

CPCs have been widely used in measuring particle emissions from stationary sources [3,4]. For example, regarding cooking emissions, ultrafine particles are significantly abundant. As these small particles account for very tiny portions of the total particles by mass but occupy great portions of total particles by number, they will significantly increase the risk of inhalation in the human respiratory tract. Hence, the traditional particle quantification method only based on mass evaluation is unable to reflect the impact of number on human environmental hazards. CPCs can provide accurate number-based measurements for these small particles originating from stationary source emissions.

In addition, CPCs have also been broadly used in quantifying particle emissions from mobile sources [5,6]. For example, in terms of vehicular emissions, China has owned the largest number of motor vehicles worldwide since 2009, and the vehicle-emitted particles have caused severe impacts on environmental quality.

DOI: 10.1201/9781003423195-5

Relevant statistic data revealed that the capacity of motor vehicles in China was over 280 million in 2013 [4,7]. Such a huge amount of motor vehicles emitted vast PM, which contributed by over 20% to the total urban PM nationally. The size range of typical nucleation PM from vehicular emissions is between 5 and 50 nm. These nucleation particles mainly include: small droplets formed by cooling and condensing processes of volatile hydrocarbons and sulfates. However, owing to the condensation and nucleation of volatile organic compounds (VOCs), which are very sensitive to external environmental conditions, they can become artifacts and bring great uncertainties when measuring the solid particles. The CPC needs to be used cautiously to measure the non-volatile particles before removing the volatile components.

Condensation CPCs are also widely used in the detection of atmospheric particles [8,9]. Large amounts of PM are generated from vehicle exhaust, coal burning in power plants, and various emission sources. The PM is directly discharged into the atmospheric environment and suspended in the air for a long time, forming a primary source of PM aerosol with a significant number concentration. Studies have shown that these particles contain trace amounts of heavy metals, toxic, VOCs, etc. The gas phase components of the PM in the primary source aerosol undergo chemical reaction denaturation in the atmospheric environment to form the secondary source PM aerosol. CPCs can provide an accurate means for measuring the primary and secondary particles by number in the atmosphere. Combined with source apportionment analysis, PM emissions from different sources can be identified and quantified.

5.2 PM MEASUREMENTS USING CPC

CPC has been widely used to measure the PM emissions originating from stationary sources, mobile sources, and ambient air. The major stationary sources of coal combustion, non-exhaust PM (brake wear, type wear, and surface wear), and cooking; mobile sources of vehicles, airplanes, and shipment have been comprehensively involved in the following discussions.

5.2.1 PM from Stationary Sources

5.2.1.1 Coal Combustion-Derived PM

Coal has been one of the most important energy sources in China. In 2015, coal still accounted for over 60% of the Chinese energy consumption system, the percentage of which is much higher than the average value of 29% worldwide [10,11]. The energy dominance of coal will not change immediately in the upcoming decade. Whereas coal combustion for power generation will continuously cause plentiful of pollutants, such as SO_2, NO_x, and $PM_{2.5}$, which are seriously harmful to the ecological environment and become a bottleneck restricting Chinese economy development sustainably [12]. It has been demonstrated that PM formation during coal combustion is a very complex physical and chemical process, and there are three major ways of combustion-derived PM formation: (1) raw coal contains a large number of mineral particles, which are directly converted into particles during combustion.

These particles are mainly formed by refractory elements, such as Si, Al, Ca, and Fe (<10 μm in diameter); (2) volatile metal elements, including alkali metals and a large number of other metals, such as Na, K, Ni, Cr, Cu, and Pb, initially release from combustion flame and then nucleate to form fine particles (<0.3 μm); (3) extrinsic minerals in the raw coal break into small components to form medium particles (0.3 μm < diameter < 10.0 μm).

PM emissions from coal-fired industrial boilers have reached 3.752 million tons in China since 2012, which accounted for 41.6% of the total PM emissions nationally. In recent years, Beijing, Tianjin, Harbin, etc., have experienced severe $PM_{2.5}$ pollution with a PM index higher than 500 μg/m^3, mainly owing to the coal-fired industry [13]. This value is far higher than the limit of 35 μg/m^3 (annual average PM2.5 concentration for Class II areas) regulated by ambient air quality standards proposed by the Chinese Ministry of Environmental Protection. Therefore, PM pollution problem originating from coal combustion has caused significant deterioration of the environmental quality. In order to control the pollution-increasing trend, the Chinese Ministry of Environmental Protection issued a standard to regulate the PM emissions from coal-fired boilers in 2014. Meanwhile, accurate measurements for the coal combustion-derived PM are necessary before control measures.

Initial research has been performed to investigate the modal distributions of coal combustion-derived PM in the early 1980s. Markowski and coworkers [14] measured the distribution of number-based PM from a 520 MW sub-bituminous coal boiler before and after dedusting processes. They observed the fine mode and coarse mode PM. Experimental results show that coal combustion generates PM with three distinct modes, which are the fine mode with a diameter below 0.5 μm, the intermediate mode with a diameter of 1–2 μm, and the coarse mode with a diameter of nearly 10 μm. Besides, Jia et al. [15] carried out coal combustion experiments under different oxygen supply levels of 21% and 31% under various temperatures of 1,410 and 1,520 K in a furnace. He concluded that coal combustion-derived PM had an ultrafine mode with a diameter less than 0.1 μm, a fine mode with a diameter between 0.1 μm and 1.0 μm, and intermediate/coarse mode with a diameter greater than 1 μm. Along with the furnace temperature increment, the ultrafine mode PM by number increased concurrently. This is owing to the formation of volatile oxides resulting in a number increase for ultrafine mode PM. Accurately measuring the number-based PM from coal combustion using CPC will help to characterize the generation and evolution of PM, which will benefit for proposing effective PM control measures for coal-fire.

5.2.1.2 Non-exhaust PM (Brake Wear, Type Wear, and Surface Wear)

PM emissions from wear and tear of vehicle parts such as brake, tire, vehicular surface, and re-suspension of dust are non-exhaust PM (Figure 5.1). Non-exhaust emissions contribute mainly to the coarse mode of PM ($PM_{2.5\text{-}10}$), while exhaust emissions contribute predominantly to fine PM (aerodynamic diameter <2.5 μm), which are prominently smaller than PM in coarse mode. As a result, the contribution of non-exhaust PM is becoming more important, although detailed information on non-exhaust PM emissions is relatively scarce. It has been shown that even

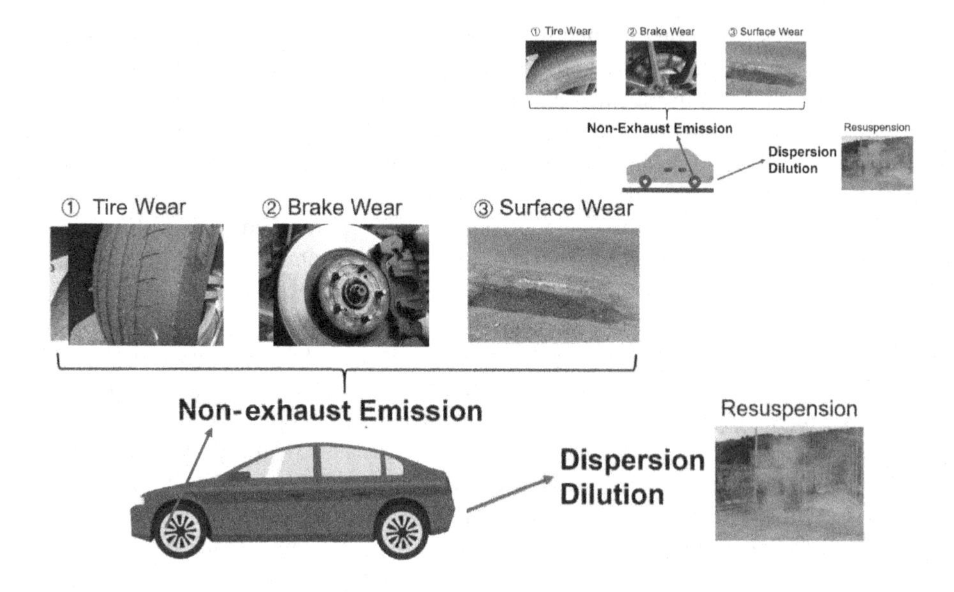

FIGURE 5.1 Non-exhausts PM emissions.

with zero tail pipe emissions, traffic will continue to contribute to fine and ultrafine particles through non-exhaust emissions, and it is estimated that nearly 90% of the total emissions from road traffic will come from non-exhaust sources by the end of the decade [16].

Non-exhaust PM comprises the various emissions that do not derive from the tail-pipe of a vehicle, including particles generated due to brake and tire wear, road surface abrasion, wear and tear/corrosion of other vehicle components such as the clutch, and re-suspension of road surface dust. The key reason to understand non-exhaust emissions is that their research is much more scarce than those of exhaust emissions, but non-exhaust emissions have great capacity to carry inherent toxicants. Amato et al. (2014) showed that the size distributions of trace metals were indicative of particle sources [16]. Keuken et al. [17] concluded that re-suspension of accumulated PM and road wear-related particles are the primary contributors to non-exhaust emissions, and tire wear and brake wear contribute to the fine and coarse fraction, respectively. Besides, non-exhaust PM was found to be a key contributor to Ba, Zn, and Pb in ultrafine particles and to Cu, Sb, Ba, Mn, and Zn in fine particles [18].

1. Type wear

   Tire wear particles are generated either by shear forces between the tire tread and the road surface and are predominantly coarse or by volatilization. Tire wear particles are reported to be generally elongated with rough surfaces based on microscopic analysis. Tire tread, a source of airborne particles, contains natural rubber copolymers such as styrene-butadiene rubber and polyisoprene rubber, and zinc (Zn) is added to tire tread as zinc oxide and organozinc compounds to facilitate the vulcanization process. Passenger car tires in EU are known to contain nearly 1% zinc oxide, and rubber, metals, and carbon black makeup typically 47%, 16.5%, and 21.5%, respectively.

For example, asphalt surfaces have been reported to cause less tire wear than concrete pavements and in Arizona, USA, the emission rate for tire wear was found to be 1.4–2 times lower for asphalt pavement compared to concrete pavement [19]. The type of tire also impacts the magnitude of tire wear emissions, and studded tires are known to cause more emissions than summer and friction tires. Unimodal (70–90 μm) and bimodal (<10 μm and 30–60 μm) peaks in the nano-size range have been reported for tire particles by number under low and high-speed conditions, respectively.

2. Brake wear

   Brake wear, including abrasion of brake lining material and brake discs, caused by grinding of brake pad constituents (coarse range particles) or volatilization and condensation of brake pad materials (fine range particles), is known to release PM directly into the atmosphere and to contribute to the trace metal concentration in airborne PM, particularly less than 10 μm [20]. Key components of brake pads include fillers, frictional additives, reinforcing fibers, and binders, and the key chemical species used include sulfides of metal, abrasives (e.g., silica), barium silicate/sulfate (particularly in brake linings), and other metallic particles (as filler material), carbon fibers and lubricant (e.g., graphite).

3. Road dust and road surface wear

   Road dust, of which crustal dust is a key component, consists of primarily coarse-sized particles derived from different sources such as traffic, industrial emissions, mineralogical dust, etc. Composition of road dust shows spatial as well as temporal variation, and it is often difficult to classify dust into crustal/re-suspended/direct emission, etc. In Monterrey (Mexico), re-suspended dust was found to be contributing nearly 20%–25% to the $PM_{2.5}$ [21]. The amount of re-suspended road dust particles depends on a number of factors including vehicle movement (particularly traffic speed), street maintenance, season, and associated meteorological parameters and speed of traffic.

5.2.1.3 Cooking-Derived PM Emissions

Recently, cooking emissions have been found to have a great impact on environmental quality and human health. Cooking activities can typically increase the indoor PM concentration by 5–90 times depending on different cooking circumstances, and the size range of the PM produced is between 0.01 and 10 μm and contain complex compositions (Figure 5.2) [22]. Moreover, cooking can generate plentiful of heavy metals (from high to low molecular weight: Cr > As > Ni > Pb > Cd), which far exceed the acceptable upper limit and can cause high carcinogenic risk. For instance, during barbecue process, carbon combustion will release large amounts of Cr, which is very harmful to human health. Besides, the carbon components including organic carbon (OC) and elemental carbon (EC) in the PM are also formed during cooking

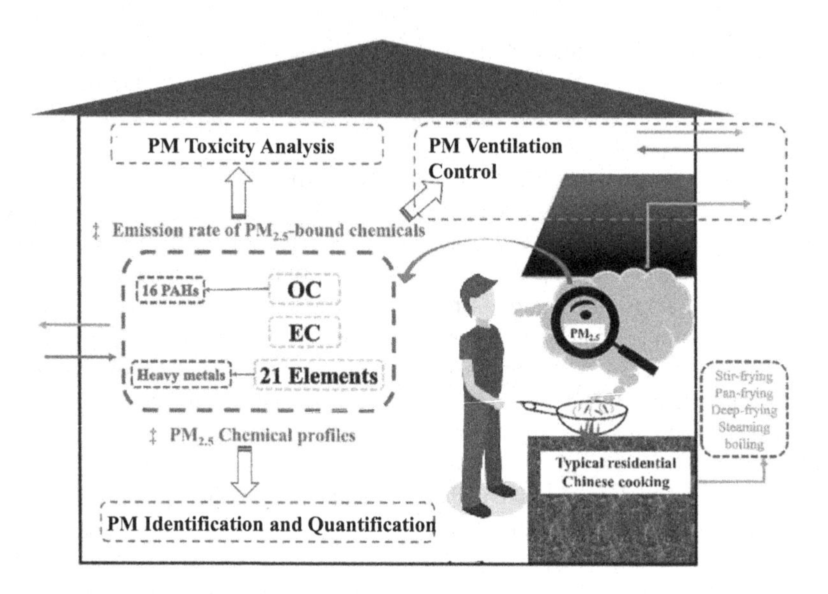

FIGURE 5.2 Cooking activities-derived PM emissions.

activities. Many previous studies have shown that OC from cooking can contribute up to 70% of the mass of $PM_{2.5}$. Nowadays, catering sources have caused equivalent PM emissions as compared to those from traffic sources in some regions. Depending on different ingredients, the OC released during cooking accounts for as high as over 50% of the fine particles, while the EC plays a much insignificant role compared to OC [23].

For the PM emissions from cooking activities, Allen et al. [24,25] studied the $PM_{2.5}$ characteristics of over 100 cooking activities and found an average concentration of 5.5 μg/m^3. Moreover, Li et al. [26] investigated the particulate size distribution and diffusion characteristics of cooking, which is thought to be helpful for figuring out particulate impact on indoor air quality. He used a scanning mobility particle sizer (SMPS) to measure the size-resolved number concentration of cooking-derived particles near and 3-m away from the cooking position, as ventilator was turned on or off, respectively. The PM is mainly fine particles with a diameter below 655 nm, and the ventilator could significantly reduce the PM concentration. Besides, as the ventilator was turned on, the number-based PM close to the cooking position decreased from 2.8×10^6 #cm^{-3} to 2.3×10^5 #cm^{-3}. As for the sampling point 3 m away from the cooking position, the total PM number concentration decreased dramatically by 65% in the absence of ventilation.

5.2.2 PM from Mobile Sources

5.2.2.1 Vehicle-Related PM Emissions

Vehicle-related particulate matter (PM) stands out as a significant source of particle emissions. These emissions demonstrate elevated concentrations and surface areas, accompanied by a complex chemical composition. This complexity profoundly impacts human health, atmospheric visibility, and the global climate. Moreover, the substantial volume of ultrafine particles has the potential for hygroscopic growth under specific atmospheric conditions, potentially contributing to the formation of haze pollution.

To exert better control over ultrafine PM emissions, the European Union took steps to regulate particulate number (PN) emissions from diesel vehicles in 2011. The implementation of the Euro 6 standard in September 2014 introduced more stringent requirements compared to the Euro 5 standard. Specifically, it lowered the PN emission limit for Gasoline Direct Injection vehicles to less than 6×10^{11} particles per kilometer. China's PM emission standards mainly draw from European standards. Since July 1, 2019, the nationwide implementation of "Emission Limits and Measurement Methods for Pollutants from Heavy-Duty Diesel Vehicles (China VI)" (GB 17691-2018) mandates that the PN emission limits for compression-ignition and dual-fuel heavy-duty diesel vehicles adhere to the requirement of 1.2×10^{12} p/kw·h.

The primary methods for measuring ultrafine PM include the electrostatic method and the condensation particle counting method. CPC has been widely applied for measurements due to its broader applicability, enhanced sensitivity, and lower detection threshold. Ground vehicles powered by gasoline engines or diesel engines are the major contributor to PM emissions. The volatile substances produced during the combustion process of motor engines tend to condense into fine particles and will cause great measurement uncertainty, which is very sensitive to the sampling conditions such as sampling point, sampling temperature, and humidity. In order to accurately analyze emitted PM, EU regulations have proposed strict standards on the sampling conditions. It is required that vehicle-related PM sampling must be operated by complying with particle measurement program. This program mainly consists of volatile particle remover, which is utilized to remove volatile components and focus on the non-volatile particles to ensure measurement reliability.

Previous major studies have performed comprehensive characterization of vehicle-related PM emissions. Li et al. [27] investigated the number concentration and size distribution of particles (size range within 10–487 nm) in a street canyon at four different heights. He found that with height increase, the number-based PM in nucleation mode decreased significantly, and its count median diameter became larger. Yang et al. [28] studied the vehicle-related PM at North Fourth Ring Road in Beijing, by comparing the normal traffic period (August 2009) and the Olympic Games period (August 2008). The purpose is to figure out how the traffic flow influence the PM profiles by number. They reported that for the normal period, three PM peaks were found at 0:00 to 4:00 when diesel-powered vehicles were abundant, at 11:00 to 13:00 when photochemical reactions were strong, and at 17:00 to 20:00 when rush hour have occurred. During the Olympic Games period, the number-based PM was dropped apparently since the traffic flow was intensively restricted. In addition, 11%–20% of the total fine organic aerosol (OA) mass concentration can be attributed to traffic emissions in Beijing, China [29].

5.2.2.2 Aviation PM Emissions

The aviation industry is currently experiencing rapid expansion with an annual increase rate of 5% (before COVID-19), which is mainly driven by fast globalization. However, the emissions from aviation aircraft have caused severe air quality problems in the airports and nearby regions. Baugh's ACI (Airport Councilcum, 2015) investigated the aviation traffic flow by analyzing 159 airports in 159 different countries in 2012 and reported that there were 0.079 billion flying activities for carrying 5.7 billion passengers and 0.093 billion cargo every year. In the next 10 years up to 2025, aviation flying activities will increase up to 0.12 billion by carrying 9 billion passengers and 0.2 billion cargo in each year [30]. As such, the continuous expansion of global aviation industry will promote the aviation networks but also cause serious environmental problems.

Civil aviation aircraft are the only pollution source at high altitude of ~11 km. At this altitude, the gas turbine combustion chamber is required to work at a relatively high temperature, high pressure, and high oil-gas ratio to ensure stable power generation, which is, therefore, causing the formation of large amounts of PM. PM emissions produced by aircraft can be divided into two categories: non-volatile particles and volatile particles. Non-volatile PMs are mainly composed of black carbon and mixed with certain metal components, etc. While volatile PM are the significant precursors for heterogeneous nucleation particles which are mainly comprised of unburned fuel and lubricating oil. These non-volatile and volatile PM become the key factors for forming wake clouds, aerosols, and even haze at high altitude, which may in turn influence the atmospheric environment and even cause regional climate change. Compared with the PM emissions emitted from ground sources (vehicle-related PM), the non-volatile PM (soot particles) originating from aircraft are featured by smaller diameter ranging from 10 to 30 nm owing to more complete combustion. These nanoparticles can act as a cloud condensation nucleus and ice nucleus (IN), which thereby affect the number and size of ice crystals in the cloud (Figure 5.3). Changes in cloud morphology and lifespan will further influence the

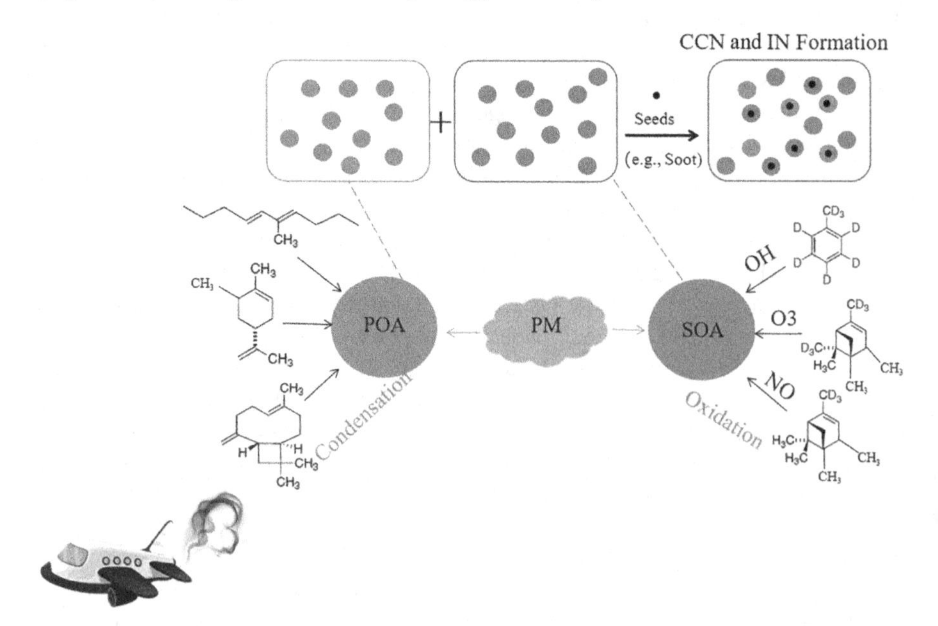

FIGURE 5.3 Aviation aircraft emitted PM emissions.

precipitation process and atmospheric radiation balance [31]. Additionally, aviation PM can be transported from upper stratosphere to lower troposphere due to advection, thereby deteriorating the pollution within troposphere layer.

In terms of the aviation PM characterization, Cheng [32] sampled the PM emissions from a C-130 Hercules military aircraft equipped with turboprop engines, the power of which was regulated at 4%, 7%, 20 %, 41%, and 100%. They reported that the average PM EI were 1.8×10^{16} #/kg·fuel, 1.4×10^{16} #/kg·fuel, 1.4×10^{16} #/kg·fuel, 1.0×10^{16} #/kg·fuel and 1.2×10^{16} #/kg·fuel at corresponding power, respectively, which exhibited to be a U-shape with power increment. Besides, Miracolo et al. [33] investigated the composition variation of PM originating from CFM-56 engine exhausts. They showed that organic components dominated the PM emissions at low power of 4% (<30 nm), while the EC became dominant at high power of 85% (50–100 nm).

5.2.2.3 Ship-Related PM Emissions

Ships are responsible for a small fraction of worldwide particle emissions, but the regional distribution must be carefully considered. The impact of shipping on environmental pollution levels is the highest along the west and east coasts of the United States, North Europe, North Pacific, and Mediterranean Sea and close to Indian coasts. The impact on population is still largely unrecognized but widely diffused. In fact, census data show that in the USA and in the European Union, about 50% of the resident population lives in coastal areas [34]. In South America and Asia (with the exception of India), 60%–75% of the population lives within 400 km from the sea. On a global scale, 23% of the world's population lives within 100 km from the shoreline, and 23–28 of the largest megalopolis, with more than 10 million inhabitants, are in coastal areas.

PM emitted by intermediate fuel oil-fueled diesel engines is a complex ensemble of different kinds of particles whose dimensions span from a few nanometers to over 20 μm. Ship emissions are composed of complex components (Figure 5.4) [34], the major particulates of which are:

1. Metals, derived from fuel impurities and wear of ship components, usually have a size between 0.2 and 10 μm.
2. Sulfates and nitrates (in minor fractions) together with associated water, usually in the micrometer size range.
3. Soot particles, largely in the submicron (<1 μm) and ultrafine (<200 nm) range, are commonly classified into two main components: EC and organic matter (OM).
4. Other components including ash, etc.

The International Maritime Organization indicated that weight fractions of sulfate and sulfur-derived particles were by far dominant, accounting for 78% in weight of the emitted particles [34]. This percentage was found to increase to 85% if ashes were included. In the same studies, the EC and OM fractions were between 2%–3% and 10%–13.5%, respectively. The toxicology of diesel particles relies on three main parameters: particle morphology, chemical–physical structure, and size. Seal and

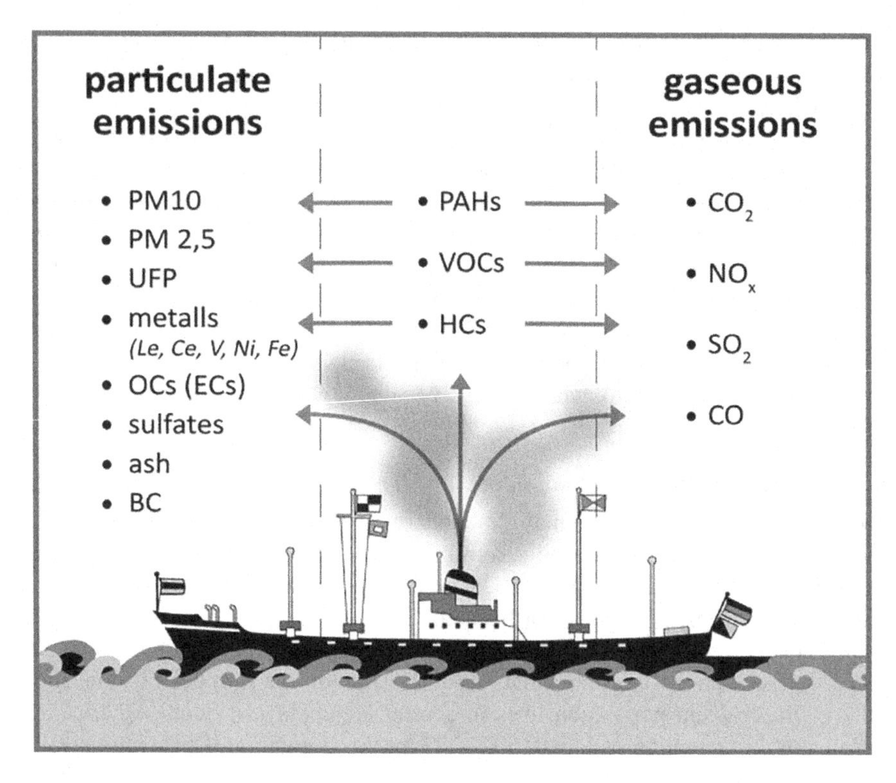

FIGURE 5.4 Ship-emitted PM along with gaseous emissions [34].

coworkers indicated that soot chemical composition gives rise to high surface reactivity and ability to cross cellular membranes [35]. Particle surface area is a key toxicological parameter since it determines the amount of available functional groups and toxic gaseous species that can be adsorbed on PM.

For ship emissions, Lack et al. [36] reported an average particle composition of 46% sulfates, 39% OM, and 15% black carbon (BC) starting from observations gathered during the Texas Air Quality Study/Gulf of Mexico Atmospheric Composition and Climate Study 2006 field campaign. In a very recent paper, Westerlund et al. [37] reported a comprehensive study on the particle emissions associated with 154 different ships. The emission factors from these ships ranged between 0.14 and 8.63×10^{16} #/kg·fuel, corresponding to about 340–5,600 mg/kg·fuel. The corresponding values of non-volatile particle emissions were $0.11–4.11 \times 10^{16}$ #/kg·fuel and 120–1,550 mg/kg·fuel. The average particle diameter was in the range of 20–40 nm.

In conclusion, sulfates and ashes are effectively reduced by the same strategies currently adopted to control sulfur dioxide emissions: the use of low sulfur fuels and conventional scrubbers proved to be effective in reducing the emissions of particles with sizes as low as 500 nm. These effects were already forsaken in the Marpol Annex VI reg. 14, where the reduction of sulfur was expected to generate a reduction of PM emission by 85% in mass.

5.2.3 PM in Ambient Air

PM is one of the main causes of air pollution in the atmosphere. The major sources of PM can be classified into natural and anthropologic types. For the natural sources, it can be forest fires, volcanic eruptions, sea dynamic processes that emit particles directly to the ground, and their secondarily forming particles. For anthropologic sources, it can mainly be fuel energy combustion, industrial production processes, transportation activities, etc. The primary PM from different sources is directly discharged into the atmospheric environment and then forms secondary aerosol via complex physical and chemical processes. These primary and secondary PM could concurrently exist in the atmosphere for a long time.

For the ambient PM, McMurry and Woo [38] conducted a 2-year study on characterizing the PM in an urban area of Atlanta and found ultrafine PM had higher concentrations during winter, working days, and rush hours. Meanwhile, nano-PM (with a smaller diameter of 3–10 nm) was more abundant during summer owing to photochemical nucleation. Liu et al. [39] investigated the PM characteristics in Los Angeles Basin and demonstrated that primary particles were mainly determined by fresh emissions from nearby highways, while secondary particles were mainly formed in the downwind area of the sources via photochemical reactions. Besides, they reported that larger particles (90–120 nm) were produced during secondary evolution. And it is observed that the ultrafine particles in urban regions were apparently higher than those in suburban regions. Besides, the PM abundance was high during the daytime but gradually decreased at night-time. Azimi et al. [40] collected ambient PM samples and assessed the spatiotemporal distribution and health risks of USEPA PAHs. This study using source apportionment analysis showed that volatilization of diesel and gasoline was the main contributor to PAHs by accounting for 37.4% and 31.1% of the entire fractions, respectively. The highest contribution rate was found in summer, while the lower contribution rates were observed in summer.

5.3 SOURCE IDENTIFICATION AND APPORTIONMENT OF ATMOSPHERIC PARTICLES

Identification of particle source types is essential for providing source-specific exposure metrics and developing effective pollution control strategies. In recent decades, aerosols have gained significant attention in scientific research due to their crucial role in climate change and human health. Numerous studies have employed source apportionment (SA) techniques using chemical composition data, utilizing receptor models such as chemical mass balance and positive matrix factorization (PMF) [41–45]. However, acquiring compositional data can often be time-consuming and expensive, involving the use of monitoring/sampling instruments for particle concentrations and various analytical techniques such as inductively coupled plasma mass spectrometry (ICP-MS), ion chromatography, gas chromatography-mass spectrometry (GC-MS), and X-ray fluorescence for chemical analysis. Meanwhile, physical characterization, such as particle number size distribution (PNSD) obtained through differential mobility analyzers (DMA) and CPC, has gained increasing importance as a key metric in determining regional lung deposition.

Hopke et al. [46] complied with 55 SA studies based on particle size distributions prior to August 1, 2021. They categorized particle sources as nucleation, traffic, heating, secondary inorganic aerosol (SIA), oxidants (such as ozone) associated sources, and others, including biomass burning and unidentified sources. The study found that nucleation and traffic were the dominant sources in terms of particle number concentrations [46]. It has been reported that analyzing data for an entire year can lead to unresolved mixed sources and difficulties in interpretation due to seasonal differences in temperature and photochemical activity. Therefore, it is necessary to apportion the sources seasonally, considering the sufficient seasonal variations in the resolved profiles [46–49].

For atmospheric particles, the PNSD ideally exhibits a multi-lognormal structure. These particles can be classified into four main modes based on their sizes: nucleation mode ($Dp < 30$ nm), Aitken mode ($30 \text{ nm} < Dp < 100$ nm), accumulation mode ($100 \text{ nm} < Dp < 1$ µm), and coarse mode ($Dp > 1$ µm) [50]. In the urban atmosphere, the particle number concentrations (PNC) are primarily dominated by the nucleation and Aitken modes. It has been reported that approximately 75% of the aerosol number is found in the size range of < 50 nm [51]. Furthermore, in five major European cities, over 80% of the particles with sizes ranging from 3 to 1,000 nm were observed in the nucleation and Aitken modes ($Dp < 100$ nm) [50,52].

5.3.1 Source Profiles of Particles Number Size Distribution

5.3.1.1 Nucleation

Newly formed particles, through nucleation, make a significant contribution to the total number of particles present in the atmosphere. The nucleation of particles can be explained by two mechanisms. The first is homogeneous nucleation, which occurs when a low-vapor-pressure species condenses without the involvement of foreign nuclei or surfaces. The second mechanism is heterogeneous nucleation, which involves the scavenging of low-vapor-pressure products on a foreign substance [50]. Nucleation events can take place in both clean environments, such as remote boreal forests, and polluted regions like industrialized agricultural areas [53]. These events are typically characterized by a shift in the peak mode toward smaller Dp due to a rapid increase in PNC within the nucleation mode. During nucleation events, PNC can increase by 3 to 10 times compared to non-event periods [54,55].

The diel peaks of nucleation events are typically linked to traffic-related temporal patterns, such as the morning and evening rush hour periods. Jeong et al. [56] observed a type of nucleation event in Rochester, which typically occurred during the morning rush hour (07:00–09:00) and exhibited a dominant particle size range of 20–100 nm. Another type of nucleation event was observed during the afternoon (peaking around 13:00) with a particle size range of 11–30 nm. These rush hour peaks are likely attributed to the nucleation of primary semi-volatile vehicular emissions as they undergo dilution and cooling [57], while afternoon peaks, which are often observed during warmer months, may be a result of new particle formation driven by high photochemical activity [46]. In European cities, the diurnal profiles of particle number counts generally correlated strongly with BC, indicating a common source in road traffic emissions. However, the increase in particle number counts during the midday, not associated with BC, can be attributed to nucleation driven by photochemistry [58].

For the seasonal occurrence of nucleation, it has been observed that regional nucleation events take place throughout all seasons, but the most intense ones are observed during spring and fall [54]. Jeong et al. [56] reported no clear seasonal variation in the afternoon nucleation occurrence frequency. However, another study noted that strong afternoon nucleation events (with PNC > 30,000 cm^{-3}) were predominantly observed in spring and summer [50].

5.3.1.2 Traffic

Hopke et al. [46] summarized two commonly reported traffic factors from the literature, which are normally associated with major number modes peaking at around 30–35 nm and 60–80 nm. The authors suggest that these two traffic factors likely represent two different forms of ignition in internal combustion vehicles. The smaller-sized particles (30~35 nm) are emitted from spark-ignition vehicles, while the larger-sized particles (60~80 nm) are emitted from diesel vehicles [46,59]. Another study proposed that the smaller-sized particle mode is associated with freshly emitted particles from traffic, whereas the larger-sized particles are formed through the coagulation of aerosols as they transport away from the sources [60]. The detailed assignment of these two traffic factors should be combined with emission characteristics of the sources, such as the involvement of diesel vehicles and the proximity of particle sampling sites to the traffic sources.

5.3.1.3 Combustion Sources

Gas fuels (such as natural gas and liquid petroleum gas), liquid fuels (such as oil), and solid fuels (such as coal, wood, and other biomass) are considered combustion sources. In general, the efficient combustion of fuels, especially natural gas, leads to the production of very small particles. Measurements of particle number size distributions and concentrations for natural gas and medium sulfur bituminous coal have shown peaks at 15–25 nm and 40–50 nm, respectively [61]. The burning of residual oil and the flaming combustion of solid fuels generate larger particles, with a distribution mode around 90–100 nm [46]. For wood boilers, the steady-state particle number size distributions exhibit a log-normal pattern, with geometric mean diameters (GMDs) between 70 and 92 nm. However, during startup periods of wood boilers, the GMD increases significantly to around 234 nm, while during shutdown conditions, it decreases substantially to 17.1 nm [62].

5.3.1.4 O_3-Rich Secondary Aerosols

In certain SA studies that include contemporaneous measurements of pollutant gases such as SO_2, NO_X, and O_3, a factor highly enriched with O_3 is identified. This factor is referred to as the O_3-rich secondary aerosol factor, as it cannot be attributed to any specific source [63,64]. The PNSD patterns in these studies reveal the presence of multiple modes, such as Aitken and accumulation modes. In some datasets, the particle number profiles are predominantly dominated by the Aitken mode, while in others, the accumulation mode particles prevail [64]. The presence of high O_3 loading, a strong correlation between O_3 and this factor, and a diel pattern characterized by early to mid-afternoon peaks suggest that this factor can be ascribed to particle growth through the condensation of both secondary organic and inorganic species [46].

5.3.1.5 Secondary Inorganic Aerosols (SIA)

A factor with a typical mode above 100 nm represents SIA, which includes ammonium sulfate and ammonium nitrate [46,64]. The seasonal patterns of this factor are affected by meteorological conditions. For instance, lower sulfate production occurs in winter due to reduced levels of photochemical activity, while lower nitrate production is observed in summer as a result of thermal dissociation of ammonium nitrate under higher temperatures. The size distribution of nitrate during winter shows a larger mode center at 250 nm, whereas the sulfate distribution during summer exhibits a mode center at approximately 100 nm [46].

5.3.1.6 Other Sources

Dust, industrial emissions, and unknown sources, which exhibit various characteristics associated with local potential sources, mostly have their primary mode in particle size distributions above 100 nm [46]. The identification of these sources is often achieved using local meteorological methods (e.g., conditional probability functions), which allow for the determination of the direction of known sources [65]. Figure 5.5 presents an overview of the average fractional contributions of the PNC sources at different locations worldwide, as derived from published SA studies using PNSD data [46].

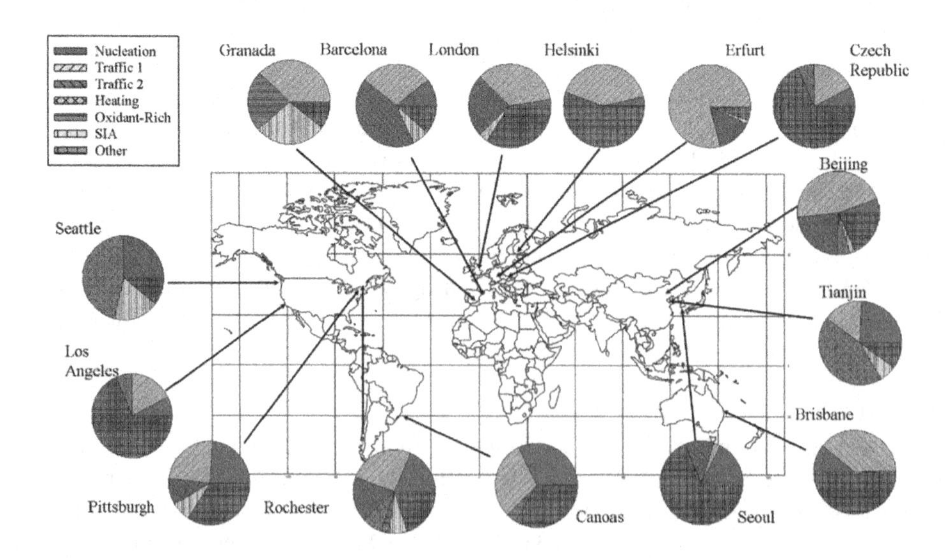

FIGURE 5.5 Average fractional contributions of the PNC sources at different locations from published source apportionment studies based on particle number size distribution data [46].

5.3.2 Source Apportionment Methods Based on Particle Number Size Distribution

Various methods can be applied to investigate the sources of ambient PM. The most commonly used SA methods based on the PNSD data are receptor models, namely PMF and principal component analysis (PCA). Both PMF and PCA can identify particle sources without prior knowledge about the sources.

PMF is a multivariate data analysis method that has been comprehensively described by Paatero and Tapper [66] and Paatero [67]. PMF can identify and apportion the sources of PM or PNSD by analyzing measurements of observed species or size bins at the receptor site. PMF is a descriptive model based on a constrained, weighted least-squares method. It does not have objective criteria for selecting the best solution. The main difference between PMF and PCA lies in their requirements. PMF necessitates non-negative component loadings and scores, while not imposing orthogonality constraints. On the other hand, PCA requires orthogonal resulting components without constraints on component loadings and scores [68]. The lack of a non-negativity requirement in PCA can result in unreasonable negative values for quantities, but this problem can be resolved by implementing a non-negativity constraint for absolute PCA scores. PCA is a rapid and convenient method for reducing the dimensionality of a dataset while preserving the information contained in the original measurements [68].

5.3.2.1 Positive Matrix Factorization (PMF)

The PMF model based on PNSD data assumes that there are p factors (representing sources, source types, or source regions) influencing the ambient aerosol X at a receptor site. It further assumes that there are linear combinations of the impacts from the p factors/sources on the observed concentrations of the various size bins. Mathematically, the ambient aerosol X (a matrix of $n \times$ observations and $m \times$ size bins) at the receptor can be represented as the product of a source matrix F and a contribution matrix G, where the elements are given by equation (5.1):

$$x_{ij} = \sum_{k=1}^{p} g_{ik} \cdot f_{jk} + e_{ij} \tag{5.1}$$

Where x_{ij} is the measured concentration of the jth species (particle size bins or chemical and gaseous species) in the ith sample. The parameter p is the total number of independent sources. The variable g_{ik} represents the contribution of the kth source to the ith sample, while f_{jk} represents the fraction of the kth source that contributes to the jth species. The parameter e_{ij} represents the residual for the jth measurement in the ith sample, which cannot be fitted by the PMF model (residuals).

In the PMF analysis, only x_{ij} is known, while the contributions (g_{ik}) and the fractions (f_{jk}) need to be estimated. The contributions and number fractions are both constrained to be non-negative. The application of PMF depends on individual uncertainty estimates for each measured data. Measured uncertainties are used for each of the x_{ij} concentrations. Samples with high uncertainty are "weighted" less in the model, meaning that these samples have less influence on the estimation of the contribution and fractions compared to samples with lower uncertainty.

Numerous studies have applied PMF for SA of ambient particles. Below are some examples, along with their sampling sites, particle size range, and instrumentation, summarized in Table 5.1.

Yue et al. [65] investigated the sources of ambient particulate matter in Erfurt, Germany, by PMF. They analyzed 29,313 hourly averaged particle size distribution measurements ranging from 0.01 to 3.0 μm. Additional data, such as gaseous pollutants (O_3, NO, NO_2, CO, and SO_2) and particle chemical species (sulfate, OC, and EC), were also used to support the interpretation of the PMF factors. The inclusion of these additional species can reduce rotational ambiguity by increasing the number of edge points [70], thus improving the accuracy of the SA results. The authors identified five sources, which explained 94% of the total NC and 99% of the total MC. The sources and their contributions were as follows: larger particles from airborne soil (1.0–2.84 μm, NC: ~0%, MC: 8%); ultrafine particles from local traffic (0.01–0.1 μm, NC: 78%, MC: 4%); secondary aerosols from local fuel combustion (mainly 0.25–0.60 μm, NC: ~0%, MC: 17%); particles from remote traffic sources (peaked at 0.1– 0.5 μm and 0.01–0.1 μm, NC: 15%, MC: 39%); secondary aerosols from multiple sources (peaked around 0.3 μm, NC: ~0%, MC: 31%). These sources were identified in all seasons, indicating their consistent contribution throughout the year.

Liu et al. [69] conducted an SA study of urban fine particle (15 nm–2.5 μm) number concentration during summertime in Beijing. To ensure the validity of the receptor model, the authors excluded days with intense nucleation events from the datasets, as nucleation events can strongly influence the number concentration of newly formed particles, which does not align with the assumption of the receptor model that the ambient concentrations are the sum of constant PNSD profiles from the contributing sources. The authors analyzed the hourly datasets using PMF and identified eight factors contributing to the particle number concentration. The factors and their characteristics are as follows: two traffic factors (major mode: ~20 nm and 40–50 nm; NC: 19.2% and 28.7%); two combustion factors associated with coal and biomass burning (major mode: ~70 nm and ~100 nm; NC: 9.4% and 20.3%); secondary nitrate factors (major mode: ~30 nm and ~150 nm; NC: 4.2%); secondary sulfate + secondary organic aerosol (SOA) (major mode: ~30 nm + ~150 nm; NC: 6.2%); fugitive dust (major mode: ~150 nm; NC: 2.9%) and regionally transported aerosol (major mode: ~200 nm; NC: 9.0%). Cumulatively, traffic and combustion aerosols accounted for 77.6% of the PNC.

Rivas et al. [70] carried out source analyses of PNSD data by PMF in urban background and traffic stations across four European cities: Barcelona, Helsinki, London, and Zurich. They concluded that the common sources in all stations were nucleation, traffic, and secondary particles. The nucleation source was further divided into photonucleation and traffic nucleation by using NOx concentrations as a proxy for traffic emissions. In addition to traffic nucleation, traffic emissions also included fresh traffic-mode particles (13–37 nm) and urban-mode particles (44–81 nm). The annual average of all traffic sources combined contributed to 71%–94% of the PNSD across all stations. Barcelona exhibited a higher contribution of photonucleation particles, ranging from 14% during the winter period to 35% during the summer, attributed

TABLE 5.1
Summary of Sampling Sites, Size Range, and Instrumentation from Published Source Apportionment Studies by PMF Using Condensation Particle Counting Technology

Sampling Site	Size range	Instrumentation	References
Erfurt	0.01–0.5 μm	DMPS consisting of a differential mobility analyzer (DMA, TSI model 3071, TSI Inc., St. Paul, MN) combined with a condensation particle counter (CPC, TSI model 3010, TSI Inc., St. Paul, MN)	[65]
	0.1–3.0 μm	Optical laser aerosol spectrometer (PMS model LAS-X, Boulder, CO)	
Institute of Atmospheric Physics, Beijing	14.5–710.5 nm	Scanning mobility particle sizer (SMPS), comprising a model TSI 3080 electrostatic classifier and a model TSI 3775 condensation particle counter (CPC)	[69]
	0.5–20 μm	TSI aerosol particle sizer (APS) 3321	
Palau Reial, Barcelona	11–478 nm	Scanning Mobility Particle Sizer Spectrometer, SMPS TSI 3936	[70]
SMEAR III Kumpula, Helsinki	6–700 nm	Differential Mobility Particle Sizer, DMPS	
Mäkelänkatu, Helsinki	6–800 nm	DMPS	
Marylebone and North Kensington, London	17–604 nm	SMPS TSI 3080	
Kaserne, Zurich	10–487 nm	SMPS TSI 3034	
Rochester	10.4–542 nm	Scanning mobility particle sizer (SMPS) consisting of a TSI model 3071 differential mobility analyzer and a TSI model 3010 condensation particle counter	[71]
University of Southern California, Los Angeles	14–760 nm	Scanning mobility particle sizer (SMPS™, TSI Model 3081) and condensation particle counter (CPC, model 3020, TSI Inc., USA)	[72]
	0.3–10 μm	Optical Particle Sizer (OPS™, Model 3330, TSI Inc., USA).	
Peking University, Beijing	3–900 nm	TDMPS (twin differential mobility particle sizer) system, composed of two parallel Hauke-type differential mobility analyzers (DMAs) that classify particles in size ranges 3–80 nm and larger than 40 nm, counting using condensation particle counter (CPC, TSI Inc., St. Paul, MN, USA) models 3025 and 3010, respectively.	[73]

to high levels of solar radiation. Biogenic emissions were identified as an important source during the summer in Helsinki, accounting for 23% during summer. Overall, the study highlights the significant contributions of traffic-related sources, nucleation, and secondary particles to the PNSD in these European cities, with variations observed depending on the location and season.

Squizzato et al. [71] conducted a study in a metropolitan area of the northeastern United States to investigate particle sources using PMF based on PNSD data ranging from 11 to 470 nm. The study covered the period from 2002 to 2016 and identified a total of eight sources contributing to the PNSD: nucleation, two traffic factors, heating, three secondary sources (O_3-rich secondary aerosol, secondary nitrate, secondary sulfate) and regionally transported aerosols. The nucleation factor was observed with a single mode in the number distribution (14–25 nm), contributing 11% (2004–2007) to 29% (2008–2010) of PNC. Two traffic factors were identified with major number modes around 30–35 nm and 58–104 nm, representing fresh-emitted particles and aged particles, respectively. These traffic factors, along with the nucleation factor, were found to be the largest contributors to PNC. The residential/commercial heating factor exhibited two modes, one smaller-sized (20–33 nm) and one coarser (111–138 nm). The O_3-rich secondary aerosol factor contributed approximately $5\% \pm 3\%$ to PNC throughout the study period. The diel patterns of this factor followed the solar radiation pattern that drives ozone formation. The secondary nitrate factor displayed three number modes with the major one between 210 and 305 nm, and the other two in the ultrafine and Aitken mode size ranges, suggesting contributions from both local and distant particle sources. The secondary sulfate factor was characterized by a major mode at around 72 nm during transition period, and exhibited various source profiles in summer periods, representing approximately 25% and 6% of the PNC, respectively. The presence of small particles (<200 nm) in this factor indicated the predominant homogeneous (gas phase) formation of sulfate. The regional transport factor explained around 4% of PNC and was mainly characterized by accumulation mode particles.

In a study conducted in central Los Angeles, PMF was used to identify and quantify major sources of ambient PNC ranging from 13 nm to 10 μm [72]. The analysis resolved six factors based on the PNSD data and additional variables such as BC, gaseous pollutants, and meteorological data. The identified factors were nucleation, traffic 1, traffic 2, urban background aerosol, secondary aerosol, and soil/road dust. Traffic 1 and traffic 2 were found to be the major contributors to PNC (60.8%–68.4%), with major number mode at 20–40 and 60–100 nm, respectively. Nucleation was another significant factor contributing to PNC (11.7%–24%), with a number mode <20 nm. Urban background aerosol (number mode: ~220 nm), secondary aerosol (number mode: ~500 nm), and soil/road dust (number mode: ≥1 μm) contributed 12.2%, 2.1%, and 1.1% of PNC, respectively. In terms of PM mass concentrations, secondary aerosols, and soil/road dust were found to have a larger mode size and were identified as the main contributors.

In a study conducted by Wang et al. [73], the air quality improvement in Beijing during the Olympic Games period in August 2008 was investigated through a series of long-term and temporary measurements. A comparison was made with the mean values of August from 2004 to 2007. The study utilized PMF to resolve four sources: local and remote traffic emissions, combustion sources, and secondary formation. The results showed significant reductions in total particle number and volume concentrations during the Olympic Games period. The total particle number and volume concentrations decreased by 41% and 35%, respectively. The four resolved sources by PMF also exhibited reductions: local traffic emissions by 47%, remote traffic emissions by 44%, combustion sources by 43%, and secondary formation by 30%. Factor 1, attributed to local traffic emissions, exhibited a peak in the PNSD at 16 nm and accounted for 25% of PNC. This factor was likely associated with traffic-induced particle nucleation, as its diurnal variation closely followed the diurnal variation trend of NOx, peaking during morning and evening rush hours. Factor 2, with a mode between 10 and 100 nm and peaking at around 50 nm, was attributed to remote traffic emissions and accounted for 29% of PNC. Factor 3, peaking at 100 nm, contributed to 33% of PNC and exhibited a diurnal variation pattern similar to BC, suggesting a main influence from combustion emissions. Factor 4, characterized by a bimodal PNSD with peaks at 30 and 200 nm, contributed to only 13% of PNC but accounted for 71% of volume concentration. This factor represented the secondary aerosol mass, supported by the high correlations between the daily patterns of secondary aerosols (such as sulfate, nitrate, ammonium, and oxygenated OA) and this factor.

Liang et al. [74] applied PMF on both PNSD data (referred as size-PMF) and organic aerosols (referred as OA-PMF) to investigate the sources of submicron particles at the northern South China Sea during a ship-based cruise campaign in summer 2018. PNSD measurements were conducted with an SMPS, while the chemical composition of submicron particles was measured using an online time of flight aerosol chemical speciation monitor. The size-PMF analysis identified four factors, including ship-affected marine primary aerosols, continent-affected marine secondary aerosols, mixed accumulation aerosols, and ship emissions, which exhibited the highest particle number concentration (44% of total PNC). OA-PMF analysis identified three factors, including low-volatility oxygenated OA (LV-OOA, 68% of the total OA mass concentration), semi-volatile OOA (SV-OOA, 21%), and hydrocarbon-like OA (HOA, 11%).

5.3.2.2 Principal Component Analysis (PCA)

Various studies have applied PCA for SA of ambient particles. Some examples are summarized in Table 5.2, including their sampling sites, particle size range, and instrumentation.

Pey et al. [75] conducted an extensive investigation into the sources of urban fine and ultrafine particles (13–800 nm) in the Western Mediterranean city of Barcelona, using a comprehensive dataset encompassing PNSD, meteorological parameters, gaseous pollutants, and $PM2_{.5}$ chemical components. By employing PCA, the researchers successfully identified seven factors or principal components that characterized the primary particle sources, including vehicle exhausts emissions, mineral dust, industrial source, sea spray, photochemically induced nucleation, regional/urban background and fuel-oil combustion. To quantify the contributions of these sources to the PNC within the 13–800 nm range, a multilinear regression analysis was employed. The results revealed that vehicle exhaust emissions emerged as the most

TABLE 5.2
Summary of Sampling Sites, Size Range, and Instrumentation from Published Source Apportionment Studies by PCA Using Condensation Particle Counting Technology

SAMPLING SITE	Size Range	Instrumentation	References
Barcelona, Spain	13–800nm	A DMPS system composed of a Condensable Particle Counter, CPC 3022, connected to a Differential Mobility Analyzer, DMA 3071	[75]
Egbert, Southern Ontario	9.3–640nm	A DMA-CPC system: a TSI 3071 differential mobility analyzer (DMA) and a TSI 3010 CPC	[68]
Hamilton, Southern Ontario	6.0–294nm	A DMA-CPC system: a TSI 3071 differential mobility analyzer (DMA) and a TSI 3025 CPC	
Simcoe, Southern Ontario	11.9–466nm	A DMA-CPC system: a TSI 3071 differential mobility analyzer (DMA) and a TSI 7610 CPC	
Leipzig	3–800nm	(a) the custom-built twin differential mobility particle sizer (TDMPS) on the basis of Vienna-type differential mobility analyzers and using dry sheath air; (b) the scanning mobility analyzer (SMPS) – similar to (a) but using a closed-loop sheath circulation and operating in scanning mode; (c) the commercial TSI 3080 SMPS.	[76]
Western Mediterranean	9–825nm	An SMPS system comprises a Differential Mobility Analyzer (DMA) and a TSI 3772 CPC	[77]

significant source, accounting for a substantial portion of the annual mean PNC (65%). Moreover, vehicle exhaust dominated PNC across all size bins, particularly in the range of 30–200nm, where it accounted for 52%–86% of the total. The second most prevalent source contributing to the annual mean PNC was the regional/urban background (24%). Together, vehicle exhaust and regional/urban background sources accounted for approximately 90% of the annual mean PNC. The remaining sources made negligible contributions, with each factor representing only 1%–3% of the total annual mean PNC. Therefore, at the urban background site in Barcelona, road traffic emerged as the primary source of ultrafine particles, while other sources may play a more significant role under specific meteorological conditions.

Chan and Mozurkewich [68] applied absolute PCA with appropriate adaptations to investigate the sources of atmospheric PM in three locations across southern Ontario. This method involved analyzing aerosol size distribution measurements, trace gas measurements, and meteorological data. Common factors at different sites were identified as photochemically formed secondary particles, regional accumulation mode particles, and boundary layer dynamic-associated trace gas variations. Additionally, site-specific factors were observed, such as emissions from local industrial sources, processed nucleation mode particles, and fine particles transported from downwind areas [68].

Costabile et al. [76] investigated the spatiotemporal variability of atmospheric particles in the size range of 3–800 nm at eight observation sites in and around Leipzig. The researchers applied PCA to aerosol, gas, and meteorological data to identify possible sources of urban aerosols. The analysis revealed several factors representing distinct particle modes and sources. One factor corresponded to particles in the accumulation mode ("droplet mode", 300–800 nm) that resulted from processes occurring in the liquid phase and long-range transport. Another factor represented particles in the accumulation mode (centered around 90–250 nm) attributed to primary emissions and subsequent aging through condensation and coagulation. Factors associated with Aitken mode (30–200 nm) and nucleation mode (5–20 nm) particles were linked to urban traffic emissions. Nucleation mode particles (3–20 nm) were considered to be photochemically formed, while aged nucleation mode particles fell within the range of 10–50 nm. The remaining factors primarily represented local sources specific to a single site or rare occurrences.

Cusack et al. [77] investigated the sources of PM_1 (particulate matter with aerodynamic diameter ≤ 1 µm) and sub-micron PNC at a regional background site in the Western Mediterranean. They employed two different techniques, PMF and PCA, to identify and characterize the sources of these particles. Using PMF, the researchers identified six major PM_1 sources, which included SOA, secondary nitrate, industrial emissions, a combination of traffic and biomass burning, fuel-oil combustion, and secondary sulfate. On the other hand, PCA analysis of sub-micron particles (9–825 nm) revealed five sources: industrial emissions combined with traffic and biomass burning; new particle formation and growth; secondary sulfate combined with fuel-oil combustion; secondary nitrate and crustal sources. Among these sources, the factor associated with new particle formation and growth accounted for the largest proportion (56%) of the total PNC, followed by the factor of industrial emissions combined with traffic and biomass burning (13% of total PNC). Other sources, such as secondary sulfate combined with fuel-oil combustion, secondary nitrate, and crustal sources, contributed 8%, 9%, and 2% of the total PNC, respectively.

5.4 NANOPARTICLE EXPOSURE ASSESSMENT AND MANAGEMENT

Fine particles pose a significant health risk as they have the ability to penetrate deep into the lungs and cause cardiovascular and respiratory diseases. In order to mitigate these risks, a comprehensive particle exposure risk assessment is essential, which involves data collection, risk analysis and evaluation, and the implementation of appropriate control measures. In the construction industry, particular attention needs to be given to the management of risks associated with the use of nanomaterials. This includes effectively managing the emission and exposure scenarios associated with these materials [78]. It is crucial to address and manage these risks at an early stage, such as during work planning or project preparation, as this enables the implementation of more effective measures to protect workers.

There are three exposure routes, including inhalation, dermal exposure, and ingestion [79]. For inhalation exposure, the proportion of particles that deposit in different regions of the respiratory system depends strongly on the particle size. For particles >300 nm, most of them deposit in the head airways due to inertial forces and gravity settling. Particles <300 nm mainly deposit through diffusion. Diffusion is the net movement of a particle caused by Brownian motion. For particles in the size range of 10–100 nm, the alveolar region shows the highest deposition fraction due to slower air movement. For dermal exposure, cuts and lacerations can facilitate dermal penetration. However, it is uncertain whether the dermal exposure route is significant for nanomaterials [80]. For oral ingestion, nanoparticles can be efficiently absorbed through the gastrointestinal tract and then translocate through the mucosal tissue into the lymphatic and circulatory systems [81]. Uptake of nanoparticles has been observed from ingestion of consumer products such as food additives and toothpaste [82].

The uncertainties of exposure assessment include determining the importance of different exposure routes, selecting appropriate occupational exposure limits, and choosing suitable instrumentation for measuring exposures [79]. Various instruments can be used for exposure assessment: a CPC or an optical particle counter can measure total number concentration; aerosol photometers can measure particle mass concentration; diffusion chargers can measure particle surface area concentration; and particle size, morphology, and composition can be assessed using devices such as SMPSs, transmission electron microscopy (TEM), scanning electron microscopy, and inductively coupled plasma mass spectrometry (ICP-MS).

For human exposure studies involving microparticles, mass is generally considered as the prime factor, while for nanosized materials, more suitable exposure metrics are employed, that is, particle size, number, surface area-to-mass ratio, geometry, crystallinity, porosity, surface functionalization, and surface charge [83]. A comparative study between the dose and three primary physicochemical parameters (particle mass, number, and surface area) is suggested for exposure analysis [84]. Some studies suggest that particle number and size are more closely associated with adverse health effects than particle mass [50]. However, there is still limited knowledge regarding which particle metric serves as the best predictor of adverse health effects.

CPCs are widely used in nanoparticle exposure assessment and management [78]. In basic exposure assessment, portable equipment, such as CPC is used to evaluate workers' exposure. For more advanced exposure assessment, other state-of-the-art instruments are also used, such as SMPS, TEM, and chemical analysis equipment (e.g., ICP-MS, GC-MS). Hand-held CPCs may vary in models, but typically, they measure particles within a number concentration range of 0 to approximately 250,000 particles/cm^3 across a size range from 10 or 20 nm to over 1.0 μm [79].

5.5 FUTURE RESEARCH IMPLICATIONS AND PROSPECTS OF CPC

For future research implications and prospects of CPC, in addition to its applications in physical-chemical characterization and SA of particles, the PNSD data obtained from CPC technologies can also contribute to health assessment models. As ambient PNC standards are implemented, there will be a need for SA and health assessment of particles based on PNSD data to serve as a scientific basis for effective air pollution control measures. Furthermore, CPC shows promise in future nanoparticle exposure assessment and management. Other potential applications include respirator fitness tests, filter tests, evaluation of mechanical filtration systems, point source monitoring, indoor air quality research, etc.

5.6 SUMMARY

Major sources of PM emissions include stationary sources such as coal combustion, non-exhaust PM (brake wear, tire wear, and surface wear), and cooking, as well as mobile sources such as vehicles, airplanes, and ships. Coal combustion is the primary source of national PM, while cooking can result in a significant amount of ultrafine PM. Other sources such as traffic, can generate different sizes of particles. The CPC is extensively utilized for measuring ambient fine particles, as well as particles emitted from stationary and mobile sources. By analyzing the particle number size distributions obtained from CPC measurements, it is possible to apportion the sources of these particles using methods such as PMF and PCA. The source profiles of major sources such as nucleation, traffic, heating, and SIA are summarized from the literature to provide insights for future SA studies. When combined with online chemical analyses of PM, this approach not only provides a fast solution for SA but also enhances the accuracy of the results through physical-chemical characterization. In addition, CPC has proven useful in nanoparticle exposure assessment and management. In the future, it holds great potential for other applications such as respirator fitness tests, filter tests, indoor air quality research, etc.

REFERENCES

[1] Kittelson DB. 1998. Engines and nanoparticles: A review. *Journal of Aerosol Science* 29(5–6):575–88.
[2] Ebi KL, McGregor G. 2008. Climate change, tropospheric ozone and particulate matter, and health impacts. *Environmental Health Perspectives* 116(11):1449–55.
[3] Chen W, Wang P, Zhang D, et al. 2020. The impact of water on particle emissions from heated cooking oil. *Aerosol Air Quality Research* 20(3):533–43.
[4] Wu Y, Zhang S, Hao J, et al. 2017. On-road vehicle emissions and their control in China: A review and outlook. *Science of the Total Environment* 574:332–49.
[5] Wang X, Caldow R, Sem GJ, et al. 2010. Evaluation of a condensation particle counter for vehicle emission measurement: Experimental procedure and effects of calibration aerosol material. *Journal of Aerosol Science* 41(3):306–18.
[6] Liu W, Kaufman SL, Osmondson BL, et al. 2006. Water-based condensation particle counters for environmental monitoring of ultrafine particles. *Journal of the Air Waste Management Association* 56(4):444–55.

[7] Wang J, Wu Q, Liu J, et al. 2019. Vehicle emission and atmospheric pollution in China: Problems, progress, and prospects. *PeerJ* 7:e6932.
[8] Sipilä M, Lehtipalo K, Kulmala M, et al. 2008. Applicability of condensation particle counters to measure atmospheric clusters. *Atmospheric Chemistry and Physics* 8(14):4049–60.
[9] Lowther SD, Jones KC, Wang X, et al. 2019. Particulate matter measurement indoors: A review of metrics, sensors, needs, and applications. *Environmental Science and Technology* 53(20):11644–56.
[10] Yuan J, Na C, Lei Q, et al. 2018. Coal use for power generation in China. *Resources, Conservation and Recycling* 129:443–53.
[11] Wang Q, Li R. 2016. Journey to burning half of global coal: Trajectory and drivers of China's coal use. *Renewable and Sustainable Energy Reviews* 58:341–6.
[12] Zhao S, Pudasainee D, Duan Y, et al. 2019. A review on mercury in coal combustion process: Content and occurrence forms in coal, transformation, sampling methods, emission and control technologies. *Progress in Energy and Combustion Science* 73:26–64.
[13] Zhai S, Jacob DJ, Wang X, et al. 2019. Fine particulate matter (PM2.5) trends in China, 2013-2018: Separating contributions from anthropogenic emissions and meteorology. *Atmospheric Chemistry and Physics* 19(16):11031–41.
[14] Markowski G, Ensor D, Hooper R, et al. 1980. A submicron aerosol mode in flue gas from a pulverized coal utility boiler. *Environmental Science and Technology* 14(11):1400–2.
[15] Wang C, Lei M, Yan W, et al. 2011. Combustion characteristics and ash formation of pulverized coal under pressurized oxy-fuel conditions. *Energy and Fuels* 25(10):4333–44.
[16] Amato F, Cassee FR, van der Gon HAD, et al. 2014. Urban air quality: The challenge of traffic non-exhaust emissions. *Journal of Hazardous Materials* 275:31–6.
[17] Keuken M, van der Gon HD, van der Valk K. 2010. Non-exhaust emissions of PM and the efficiency of emission reduction by road sweeping and washing in the Netherlands. *Science of the Total Environment* 408(20):4591–9.
[18] Pérez N, Pey J, Cusack M, et al. 2010. Variability of particle number, black carbon, and PM10, PM2. 5, and PM1 levels and speciation: Influence of road traffic emissions on urban air quality. *Aerosol Science and Technology* 44(7):487–99.
[19] Birmili W, Allen AG, Bary F, et al. 2006. Trace metal concentrations and water solubility in size-fractionated atmospheric particles and influence of road traffic. *Environmental Science and Technology* 40(4):1144–53.
[20] Garg BD, Cadle SH, Mulawa PA, et al. 2000. Brake wear particulate matter emissions. *Environmental Science and Technology* 34(21):4463–9.
[21] Mancilla Y, Mendoza A. 2012. A tunnel study to characterize PM2. 5 emissions from gasoline-powered vehicles in Monterrey, Mexico. *Atmospheric Environment* 59:449–60.
[22] Naseri M, Rahmanikhah Z, Beiygloo V, et al. 2018. Effects of two cooking methods on the concentrations of some heavy metals (cadmium, lead, chromium, nickel and cobalt) in some rice brands available in Iranian market. *Journal of Chemical Health Risks* 4(2):65–72.
[23] Lin P, Gao J, He W, et al. 2021. Estimation of commercial cooking emissions in real-world operation: Particulate and gaseous emission factors, activity influencing and modelling. *Environmental Pollution* 289:117847.
[24] Patel S, Sankhyan S, Boedicker EK, et al. 2020. Indoor particulate matter during HOMEChem: Concentrations, size distributions, and exposures. *Environmental Science and Technology* 54(12):7107–16.
[25] Arata C, Misztal PK, Tian Y, et al. 2021. Volatile organic compound emissions during HOMEChem. *Indoor Air* 31(6):2099–117.
[26] Li S, Xu J, Mo S, et al. 2017. 模拟烹饪油烟的粒径分布与扩散. 环境科学 38(1):33–40.

[27] Li X, Wang J, Tu X, et al. 2007. Vertical variations of particle number concentration and size distribution in a street canyon in Shanghai, China. *Science of the Total Environment* 378(3):306–16.
[28] Zhou Y, Wu Y, Yang L, et al. 2010. The impact of transportation control measures on emission reductions during the 2008 Olympic Games in Beijing, China. *Atmospheric Environment* 44(3):285–93.
[29] Deng W, Hu Q, Liu T, et al. 2017. Primary particulate emissions and secondary organic aerosol (SOA) formation from idling diesel vehicle exhaust in China. *Science of the Total Environment* 593:462–9.
[30] Masiol M, Harrison RM. 2014. Aircraft engine exhaust emissions and other airport-related contributions to ambient air pollution: A review. *Atmospheric Environment* 95:409–55.
[31] Koehler KA, DeMott PJ, Kreidenweis SM, et al. 2009. Cloud condensation nuclei and ice nucleation activity of hydrophobic and hydrophilic soot particles. *Physical Chemistry Chemical Physics* 11(36):7906–20.
[32] Cheng MD. 2009. *A Comprehensive Program for Measurements of Military Aircraft Emissions*. Oak Ridge National Lab, Oak Ridge, TN.
[33] Miracolo MA, Drozd GT, Jathar SH, et al. 2012. Fuel composition and secondary organic aerosol formation: Gas-turbine exhaust and alternative aviation fuels. *Environmental Science and Technology* 46(15):8493–501.
[34] Di Natale F, Carotenuto C. 2015. Particulate matter in marine diesel engines exhausts: Emissions and control strategies. *Transportation Research Part D: Transport Environmental science* 40:166–91.
[35] Seal S, Jeyaranjan A, Neal CJ, et al. 2020. Engineered defects in cerium oxides: Tuning chemical reactivity for biomedical, environmental, & energy applications. *Nanoscale* 12(13):6879–99.
[36] Lack DA, Corbett JJ, Onasch T, et al. 2009. Particulate emissions from commercial shipping: Chemical, physical, and optical properties. *Journal of Geophysical Research: Atmospheres* 114:D00F04.
[37] Westerlund J, Hallquist M, Hallquist ÅM. 2015. Characterization of fleet emissions from ships through multi-individual determination of size-resolved particle emissions in a coastal area. *Atmospheric Environment* 112:159–66.
[38] McMurry PH, Woo KS. 2002. Size distributions of 3-100nm urban Atlanta aerosols: Measurement and observations. *Journal of Aerosol Medicine* 15(2):169–78.
[39] Liu H, Qi L, Liang C, et al. 2020. How aging process changes characteristics of vehicle emissions? A review. *Critical Reviews in Environmental Science and Technology* 50(17):1796–828.
[40] Azimi-Yancheshmeh R, Moeinaddini M, Feiznia S, et al. 2021. Seasonal and spatial variations in atmospheric PM2. 5-bound PAHs in Karaj city, Iran: Sources, distributions, and health risks. *Sustainable Cities and Society* 72:103020.
[41] Srivastava D, Xu J, Liu D, et al. 2021. Insight into PM2.5 sources by applying Positive Matrix factorization (PMF) at urban and rural sites of Beijing. *Atmospheric Chemistry and Physics* 21(19):14703–24.
[42] Xu J, Liu D, Wu X, et al. 2021. Source apportionment of fine organic carbon at an urban site of Beijing using a chemical mass balance model. *Atmospheric Chemistry and Physics* 21(9):7321–41.
[43] Chen P, Wang T, Hu X, et al. 2015. Chemical mass balance source apportionment of size-fractionated particulate matter in Nanjing, China. *Aerosol Air Quality Researh* 15(5):1855–67.
[44] Miranda RM, de Fatima Andrade M, Dutra Ribeiro FN, et al. 2018. Source apportionment of fine particulate matter by positive matrix factorization in the metropolitan area of São Paulo, *Brazil. Journal of Cleaner Production* 202:253–63.

[45] Ikemori F, Uranishi K, Asakawa D, et al. 2021. Source apportionment in PM2.5 in central Japan using positive matrix factorization focusing on small-scale local biomass burning. *Atmospheric Pollution Research* 12(3):162–72.
[46] Hopke PK, Feng Y, Dai Q. 2022. Source apportionment of particle number concentrations: A global review. *Science of the Total Environment* 819:153104.
[47] Zhou L, Kim E, Hopke PK, et al. 2004. Advanced factor analysis on Pittsburgh particle size-distribution data. *Aerosol Science and Technology* 38(Suppl. 1):118–32.
[48] Zhou L, Kim E, Hopke PK, et al. 2005. Mining airborne particulate size distribution data by positive matrix factorization. *Journal of Geophysical Research: Atmospheres* 110(D7):D07S19-n.
[49] Zhou L, Hopke PK, Stanier CO, et al. 2005. Investigation of the relationship between chemical composition and size distribution of airborne particles by partial least squares and positive matrix factorization. *Journal of Geophysical Research: Atmospheres* 110(D7):D07S18.
[50] Vu TV, Delgado-Saborit JM, Harrison RM. 2015. Review: Particle number size distributions from seven major sources and implications for source apportionment studies. *Atmospheric Environment* 122:114–32.
[51] Stanier CO, Khlystov AY, Pandis SN. 2004. Ambient aerosol size distributions and number concentrations measured during the Pittsburgh Air Quality Study (PAQS). *Atmospheric Environment* 38(20):3275–84.
[52] von Bismarck-Osten C, Birmili W, Ketzel M, et al. 2013. Characterization of parameters influencing the spatio-temporal variability of urban particle number size distributions in four European cities. *Atmospheric Environment* 77:415–29.
[53] Kulmala M, Vehkamäki H, Petäjä T, et al. 2004. Formation and growth rates of ultrafine atmospheric particles: A review of observations. *Journal of Aerosol Science* 35(2):143–76.
[54] Stanier CO, Khlystov AY, Pandis SN. 2004. Nucleation events during the Pittsburgh Air Quality Study: Description and relation to key meteorological, gas phase, and aerosol parameters. *Aerosol Science and Technology* 38(Suppl. 1):253–64.
[55] Wehner B, Wiedensohler A, Tuch TM, et al. 2004. Variability of the aerosol number size distribution in Beijing, China: New particle formation, dust storms, and high continental background. *Geophysical Research Letters* 31(22):L22108-1.
[56] Jeong CH, Hopke PK, Chalupa D, et al. 2004. Characteristics of nucleation and growth events of ultrafine particles measured in Rochester, NY. *Environmental Science and Technology* 38(7):1933–40.
[57] Harrison RM, Beddows DCS, Dall'Osto M. 2011. PMF analysis of wide-range particle size spectra collected on a major highway. *Environmental Science and Technology* 45(13):5522–8.
[58] Reche C, Querol X, Alastuey A, et al. 2011. New considerations for PM, Black Carbon and particle number concentration for air quality monitoring across different European cities. *Atmospheric Chemistry and Physics* 11(13):6207–27.
[59] Johnson JP, Kittelson DB, Watts WF. 2005. Source apportionment of diesel and spark ignition exhaust aerosol using on-road data from the Minneapolis metropolitan area. *Atmospheric Environment* 39(11):2111–21.
[60] Zhu Y, Hinds WC, Kim S, et al. 2002. Study of ultrafine particles near a major highway with heavy-duty diesel traffic. *Atmospheric Environment* 36(27):4323–35.
[61] Chang MCO, Chow JC, Watson JG, et al. 2004. Measurement of ultrafine particle size distributions from coal-, oil-, and gas-fired stationary combustion sources. *Journal of the Air & Waste Management Association* 54(12):1494–505.
[62] Chandrasekaran SR, Laing JR, Holsen TM, et al. 2011. Emission characterization and efficiency measurements of high-efficiency wood boilers. *Energy & Fuels* 25(11):5015–21.

[63] Ogulei D, Hopke PK, Chalupa DC, et al. 2007. Modeling source contributions to submicron particle number concentrations measured in Rochester, New York. *Aerosol Sci Technol* 41(2):179–201.
[64] Kasumba J, Hopke PK, Chalupa DC, et al. 2009. Comparison of sources of submicron particle number concentrations measured at two sites in Rochester, NY. *Science of The Total Environment* 407(18):5071–84.
[65] Yue W, Stölzel M, Cyrys J, et al. 2008. Source apportionment of ambient fine particle size distribution using positive matrix factorization in Erfurt, Germany. *Science of the Total Environment* 398(1):133–44.
[66] Paatero P, Tapper U. 1994. Positive matrix factorization: A non-negative factor model with optimal utilization of error estimates of data values. *Environmetrics* 5(2):111–26.
[67] Paatero P. 1997. Least squares formulation of robust non-negative factor analysis. *Chemometrics and Intelligent Laboratory Systems* 37(1):23–35.
[68] Chan TW, Mozurkewich M. 2007. Application of absolute principal component analysis to size distribution data: Identification of particle origins. *Atmospheric Chemistry and Physics* 7(3):887–97.
[69] Liu ZR, Hu B, Liu Q, et al. 2014. Source apportionment of urban fine particle number concentration during summertime in Beijing. *Atmospheric Environment* 96:359–69.
[70] Rivas I, Beddows DCS, Amato F, et al. 2020. Source apportionment of particle number size distribution in urban background and traffic stations in four European cities. *Environment International* 135:105345.
[71] Squizzato S, Masiol M, Emami F, et al. 2019. Long-term changes of source apportioned particle number concentrations in a metropolitan area of the northeastern United States. *Atmosphere* 10(1):27.
[72] Sowlat MH, Hasheminassab S, Sioutas C. 2016. Source apportionment of ambient particle number concentrations in central Los Angeles using positive matrix factorization (PMF). *Atmospheric Chemistry and Physics* 16(8):4849–66.
[73] Wang ZB, Hu M, Wu ZJ, et al. 2013. Long-term measurements of particle number size distributions and the relationships with air mass history and source apportionment in the summer of Beijing. *Atmospheric Chemistry and Physics* 13(20):10159–70.
[74] Liang B, Cai M, Sun Q, et al. 2021. Source apportionment of marine atmospheric aerosols in northern South China Sea during summertime 2018. *Environmental Pollution* 289:117948.
[75] Pey J, Querol X, Alastuey A, et al. 2009. Source apportionment of urban fine and ultra-fine particle number concentration in a Western Mediterranean city. *Atmospheric Environment* 43(29):4407–15.
[76] Costabile F, Birmili W, Klose S, et al. 2009. Spatio-temporal variability and principal components of the particle number size distribution in an urban atmosphere. *Atmospheric Chemistry and Physics* 9(9):3163–95.
[77] Cusack M, Pérez N, Pey J, et al. 2013. Source apportionment of fine PM and sub-micron particle number concentrations at a regional background site in the western Mediterranean: A 2.5 year study. *Atmospheric Chemistry and Physics* 13(10):5173–87.
[78] Silva F, Arezes P, Swuste P. 29 - Risk management: Controlling occupational exposure to nanoparticles in construction. In: Pacheco-Torgal F, Diamanti MV, Nazari A, Granqvist CG, Pruna A, Amirkhanian S, editors. *Nanotechnology in Eco-efficient Construction* (Second Edition). Woodhead Publishing; 2019, pp. 755–84.
[79] Peters TM, Ramachandran G, Park JY, et al. Chapter 2 - Assessing and managing exposures to nanomaterials in the workplace. In: Ramachandran G, editor *Assessing Nanoparticle Risks to Human Health* (Second Edition). Oxford: William Andrew Publishing; 2016, pp. 21–44.
[80] Labouta HI, Schneider M. 2013. Interaction of inorganic nanoparticles with the skin barrier: Current status and critical review. *Nanomedicine: Nanotechnology, Biology, and Medicine* 9(1):39–54.

[81] Moghimi SM, Hunter AC, Murray JC. 2001. Long-circulating and target-specific nanoparticles: Theory to practice. *Pharmacological Reviews* 53(2):283–318.
[82] Fröhlich E, Roblegg E. 2012. Models for oral uptake of nanoparticles in consumer products. *Toxicology* 291(1–3):10–7.
[83] Oberdörster G, Maynard A, Donaldson K, et al. 2005. Principles for characterizing the potential human health effects from exposure to nanomaterials: Elements of a screening strategy. *Particle and Fibre Toxicology* 2:8.
[84] Sharma S, Mehta SK, Parmar A, et al. Chapter 18 - Understanding toxicity of nanomaterials in the environment: Crucial tread for controlling the production, processing, and assessing the risk. In: Hussain CM, editor *Nanomaterials in Chromatography*. Elsevier; 2018, pp. 467–500.

Index

Printed in the United States
by Baker & Taylor Publisher Services